The Dilbit Disaster

Inside The Biggest Oil Spill You've Never Heard Of

ELIZABETH MCGOWAN

LISA SONG

DAVID HASEMYER

Copyright © 2012 InsideClimate News

All rights reserved.

ISBN: 1539009599
ISBN-13: 978-1539009597

OTHER TITLES

Exxon: The Road Not Taken (2015)
Meltdown: Terror at the Top of the World (2014)
Keystone & Beyond (2014)
Big Oil & Bad Air (2014)
Bloomberg's Hidden Legacy (2013)
Clean Break (2012)

CONTENTS

Why We Went After the Story and How We Got It   Pg. 1

The Michigan Investigation

1. Like Stepping On Chewing Gum   Pg. 5
2. You Can't Just Evacuate an Entire County   Pg. 16
3. A Moving Target on the River Bottom   Pg. 25
4. Epilogue: Cleanup, Consequences and Lives Changed in the Dilbit Disaster   Pg. 36

Follow-Up Reports

5. EPA Worries Dilbit Still a Threat to Kalamazoo River More Than Two Years After Spill   Pg. 44
6. Angry Michigan Residents Fight Uneven Battle Against Pipeline Project on Their Land   Pg. 49
7. Few Oil Pipeline Spills Detected By Much-Touted Technology   Pg. 58
8. Keystone XL Would Not Use Most Advanced Spill Protection Technology   Pg. 64
9. Little Oversight For Enbridge Pipeline Route That Skirts Lake Michigan   Pg. 71
10. New Pipeline Safety Regulations Won't Apply to Keystone XL   Pg. 79
11. Extra: A Dilbit Primer   Pg. 86

WHY WE WENT AFTER THE STORY AND HOW WE GOT IT

In 2009, InsideClimate News began reporting on a little-known pipeline project called the Keystone XL, which would originate in the tar sands region of Alberta, Canada, cross into the United States and travel 1,702 miles to the Texas Gulf Coast.

Aside from the people who lived along the proposed route, most Americans had never heard of the project.

But environmentalists had already begun working to stop the Keystone XL, even though their chances of success seemed small. Two similar pipelines had been approved by the U.S. State Department, and this one was expected to be a shoo-in, too.

The Keystone XL would carry an unconventional and low-grade crude known as bitumen. Because it is thick like peanut butter, bitumen can't flow out of wells as conventional oil does. Instead, it is mined from the earth's surface or extracted by injecting steam deep into the ground. Then it must be thinned with a cocktail of hydrocarbon liquids, including benzene, so it can be moved through pipelines. At that point it is called diluted bitumen, or dilbit.

These processes—plus the emissions produced by refining the bitumen and then burning the resulting products—create about 20 percent more greenhouse gases than extracting and refining conventional oil. Approving the Keystone XL pipeline would encourage development of the tar sands and accelerate global warming, environmentalists warned.

Along the proposed route of the pipeline, more personal and immediate worries weighed on ranchers and farmers, many of them descended from

Dust Bowl survivors, as InsideClimate News reporters Elizabeth McGowan and Lisa Song discovered when they visited Nebraska and South Dakota in 2010 and 2011.

What would happen if a pipeline accident spilled oil onto their land, the ranchers asked—land that in some cases can be reached only by dirt roads? And what was tar sands oil, anyway? (They hadn't yet learned the term "dilbit.") What would happen if it contaminated the Ogallala aquifer, which supplies their drinking and irrigation water and in some areas lies just a few feet below the surface of their land? They couldn't survive without that water, they said again and again.

Lisa began searching for answers. A trained scientist with a degree from MIT, she expected to find research papers and experts who could help her. But she found no peer-reviewed research about how dilbit behaves in water, and the oil industry experts she talked with were of little help. Dilbit wasn't imported into the United States until about 1999, and then only in relatively small quantities. Nobody seemed to know much about it.

A picture began to emerge, of people who were trying to protect their land for future generations but had been given no voice in the debate over a pipeline that could carry almost 35 million gallons of dilbit across their properties each day. State and federal regulators and politicians hadn't even answered some of their most basic questions.

Inspired by their plight, we set out to answer those questions ourselves. With just four reporters and a small budget, it was by far the most ambitious project our five-year-old organization had undertaken. But it seemed right, almost inevitable. If we didn't find some answers, who would? The ranchers' story was also part of a broader issue that InsideClimate News had been covering for years, about the nation's energy future and its attempts to address global climate change.

We began by sending Elizabeth McGowan to the site of a July 25, 2010 accident on an older pipeline that had poured more than a million gallons of dilbit into Michigan's Kalamazoo River. With no scientific data to guide us, that spill became our laboratory.

Elizabeth arrived in Marshall, Mich. in November 2011 and was shocked to find that about 36 miles of the Kalamazoo River were still closed and that the cleanup was far from over. Yet somehow this environmental disaster—one of the biggest inland oil spills in U.S. history and the nation's first major dilbit spill—had gone unnoticed by most of the news media.

Why had the pipeline ruptured? Why was the cleanup so difficult? Like the farmers' questions in Nebraska and South Dakota, these questions, too, were unanswered

Elizabeth interviewed people who were evacuated from homes near the ruptured pipeline, people who rushed to save birds and other wildlife from the gush of oil, people who were sickened by the acrid stench that lingered for days over the town. She interviewed local, state and federal officials, including leaders of the EPA team sent to supervise the massive cleanup operation. Meanwhile, back in Boston, Lisa scrutinized thousands of pages of documents and government pipeline records, searching for clues about what had happened, and why.

"The Dilbit Disaster: The Biggest Oil Spill You've Never Heard Of" is the result of their work. It's the dramatic story of people who woke up one morning to find their lives turned upside down by a pipeline many of them didn't know existed, and of a federal bureaucracy struggling to cope with a disaster that had slipped through the cracks of its outdated rules and regulations.

The ruptured pipeline had a long history of corrosion problems, we learned, including a defect near the rupture point that had been documented at least three times but was never fixed. We also discovered that federal regulations are so weak that the Canadian pipeline operator, Enbridge, Inc., wasn't even required to tell first responders what the pipeline was carrying when it ruptured. The EPA didn't know it was dealing with dilbit until more than a week into the cleanup.

Most important, Lisa and Elizabeth proved—through the interviews they conducted and the documents they scrutinized—that dilbit reacts very differently from conventional oil when it spills into water. After the pipeline ruptured, the liquid chemicals in the dilbit began evaporating, leaving behind the heavy bitumen to sink to the river bottom. Benzene, a known carcinogen, drifted into the air, but officials didn't have enough information to fully evaluate its health risks and waited four days to recommend an evacuation.

Many of the cleanup techniques that work well for conventional oil—which floats on water—proved ineffective on this spill. The EPA and Enbridge had to figure out how to find and measure the sunken oil. They're still figuring out how to remove it even today, almost three years after the pipeline ruptured.

In March 2013, the EPA ordered Enbridge to dredge several miles of the river where oil is still accumulating. The goal is to keep the oil from moving into parts of the river that have so far been protected. At $820 million and counting, it is the most expensive oil pipeline spill in U.S. history.

After "The Dilbit Disaster" was published in July 2012, Lisa, Elizabeth and InsideClimate News reporter David Hasemyer wrote dozens more articles. Based on that work, we now know that the public is more likely to detect pipeline spills than the technology used on most of the nation's pipelines. We also know that more effective safety technology is available, but federal regulations don't require pipeline operators to use it and it isn't planned for the Keystone XL or for the thousands of miles of other new and repurposed pipelines planned for the United States.

In April, Lisa, Elizabeth and David won journalism's most prestigious honor, the 2013 Pulitzer Prize for National Reporting, for their work. (The articles that won the prize are collected in this book.) They also won the 2013 Aronson Award for Social Justice. But these honors don't mean that our reporting about dilbit and the nation's poorly regulated pipeline system has ended.

Production in Alberta's tar sands region continues to boom, with most of the bitumen headed to U.S. refineries. About 1.5 million barrels of dilbit now cross the border every day, a number that could rise to 5 million barrels a day over the next 15 years.

On March 29 another dilbit spill occurred, this one in a cluster of tidy brick homes in Mayflower, Ark. Lisa visited the town and brought home more stories about people whose lives had been upended by a pipeline accident. It was another dilbit disaster and we remembered, again, the people in Nebraska and South Dakota whose questions inspired our work.

Those ranchers and farmers are still waiting for the Obama administration to decide whether to allow the Keystone XL pipeline to be buried on their hard-won land. This book answers many of the questions they asked almost two years ago. But it also raises many more that need to be addressed about the place of dilbit pipelines in the nation's energy future.

Susan White, Executive Editor
InsideClimate News
May 2013.

The Michigan Investigation

1. LIKE STEPPING ON CHEWING GUM

MARSHALL, Mich.—An acrid stench had already enveloped John LaForge's five-bedroom house when he opened the door just after 6 a.m. on July 26, 2010. By the time the building contractor hurried the few feet to the refuge of his Dodge Ram pickup, his throat was stinging and his head was throbbing.

LaForge was at work excavating a basement when his wife called a couple of hours later. The odor had become even more sickening, Lorraine told him. And a fire truck was parked in front of their house, where Talmadge Creek rippled toward the Kalamazoo River.

LaForge headed home. By the time he arrived, the stink was so intense that he could barely keep his breakfast down.

Something else was wrong, too.

Water from the usually tame creek had inundated his yard, the way it often did after heavy rains. But this time a black goo coated swaths of his golf course-green grass. It stopped just 10 feet from the metal cap that marked his drinking water well. Walking on the tarry mess was like stepping on chewing gum.

LaForge said he was stooped over the creek, looking for the source of the gunk, when two men in a white truck marked Enbridge pulled up just before 10 a.m. One rushed to LaForge's open front door and disappeared inside with an air-monitoring instrument.

The man emerged less than a minute later, and uttered the words that still haunt LaForge today: It's not safe to be here. You're going to have to leave your house. Now.

John and Lorraine LaForge, their grown daughter and one of the three grandchildren living with them at the time piled into the pickup and their minivan as fast as they could, given Lorraine's health problems. They didn't pause to grab toys for the baby or extra clothes for the two children at preschool. They didn't even lock up the house.

Within a half hour, they had checked into two rooms at a Holiday Inn Express, which the family of six would call home for the next 61 days.

Their lives had been turned upside down by a pipeline accident that triggered the first major spill of Canadian diluted bitumen in a U.S. river. Diluted bitumen is the same type of oil that could someday be carried by the much-debated Keystone XL pipeline. If that project is approved, the section that runs through Nebraska will cross the Ogallala aquifer, which supplies drinking water for eight states as well as 30 percent of the nation's irrigation water.

"People don't realize how your life can change overnight," LaForge told an InsideClimate News reporter as they drove slowly past his empty house in November 2011. "It has been devastating."

The spill happened in Marshall, a community of 7,400 in southwestern Michigan. At least 1 million gallons of oil blackened more than two miles of Talmadge Creek and almost 36 miles of the Kalamazoo River, and oil is still showing up 23 months later, as the cleanup continues. About 150 families have been permanently relocated and most of the tainted stretch of river between Marshall and Kalamazoo remained closed to the public until June 21, 2012.

The accident was triggered by a six-and-a-half foot tear in 6B, a 30-inch carbon steel pipeline operated by Enbridge Energy Partners, the U.S. branch of Enbridge Inc., Canada's largest transporter of crude oil. With Enbridge's costs already totaling more than $765 million, it is the most expensive oil pipeline spill since the U.S. government began keeping records in 1968. An independent federal agency, the National Transportation Safety Board, is investigating the accident, and the U.S. Environmental Protection Agency has launched criminal and civil probes.

Despite the scope of the damage, the Enbridge spill hasn't attracted much national attention, perhaps because it occurred while oil was still spewing from BP's Macondo well in the Gulf of Mexico, which exploded three months earlier. Early reports about the Enbridge spill also downplayed its seriousness. Just about everybody, including the EPA officials who rushed to Marshall, expected the mess to be cleaned up in a couple of months.

What the EPA didn't know then, however, was that 6B was carrying bitumen, the dirtiest, stickiest oil on the market.

Bitumen is so thick—about the consistency of peanut butter—that it doesn't flow from a well like the crude oil found in most of the nation's pipelines. Instead the tarry resin is either steamed or strip-mined from sandy soil. Then it is thinned with large quantities of liquid chemicals so it can be pumped through pipelines. These diluents usually include benzene, a known human carcinogen. At this point it becomes diluted bitumen, or dilbit.

Some environmental organizations say dilbit is so acidic and abrasive that it's more likely to corrode and weaken pipes than conventional oil. The oil industry disputes that hypothesis. It says dilbit is no different from conventional crude.

No independent scientific research has been done to determine who is right. But a seven-month investigation of the Enbridge spill by InsideClimate News has revealed one fact neither side disputes: The cleanup of the Kalamazoo River dilbit spill was unlike any cleanup the EPA had ever tackled before.

Instead of remaining on top of the water, as most conventional crude oil does, the bitumen gradually sank to the river's bottom, where normal cleanup techniques and equipment were of little use. Meanwhile, the benzene and other chemicals that had been added to liquefy the bitumen evaporated into the air.

InsideClimate News also learned that federal and local officials didn't discover until more than a week after the spill that 6B was carrying dilbit, not conventional oil. Federal regulations do not require pipeline operators to disclose that information. And Enbridge officials did not volunteer it.

Mark Durno, an EPA deputy incident commander who is still involved in the cleanup in Marshall, is among those who were surprised by what they found.

"Submerged oil is what makes this thing more unique than even the Gulf of Mexico situation," Durno told InsideClimate News. "Yes, that was huge—but they knew the beast they were dealing with. This experience was brand new for us. It would have been brand new for anyone in the United States."

Jim Rutherford, the public health officer for Michigan's Calhoun County, said he had "no idea what I was driving into," when he rushed to Marshall the day 6B ruptured.

"Enbridge was caught off guard initially, much like all of us were," Rutherford said in an interview. "We just weren't ready for anything of this magnitude. ... We didn't even know the nature of the type of crude."

Pipeline 6B was built in 1969 and is 293 miles long. It is part of Enbridge's 1,900-mile Lakehead system, which transports Canadian oil to major refining centers in the Great Lakes region, the Midwest and Ontario.

In 1999 Enbridge was among the first pipeline operators to bring Canadian dilbit into the United States. Every day, more than 11.3 million gallons of Canadian oil is transferred to 6B at Enbridge's terminal in

Griffith, Ind., and pumped across southern Michigan, to Sarnia, in the province of Ontario, Canada. From Sarnia, it is transferred to lines that connect to refineries near Detroit and surrounding markets. On the day of the spill, 6B was moving a mixture of two types of dilbit—about one-quarter Western Canadian Select and three-quarters Cold Lake.

The federal agency responsible for regulating interstate pipelines is the Pipeline and Hazardous Materials Safety Administration (PHMSA), a perennially underfunded and understaffed division of the U.S. Department of Transportation. For the most part the agency relies on pipeline operators to monitor their pipelines and self-report any problems. One of the biggest concerns is corrosion, which can lead to spills or leaks if the corroded areas aren't patched or replaced.

When corrosion rises above a certain threshold, PHMSA requires that it be repaired within 180 days. But the rules are flexible, and companies can easily negotiate for more time.

Records show that 6B had a history of corrosion problems.

In 2008, Enbridge identified 140 corrosion defects on 6B as serious enough to fall into the 180-day category. But the company repaired just 26 of them during that period.

In 2009, Enbridge self-reported a separate set of 250 defects to PHMSA. The company fixed only 35 of them within 180 days.

Instead of immediately addressing the 329 defects that now remained, Enbridge got a one-year extension from PHMSA by exercising its legal option to reduce pumping pressure on 6B while it decided whether to repair or replace the line.

A defect on 6B near John LaForge's house, where the pipeline eventually ruptured, didn't appear on any of the 180-day repair lists.

That defect, at mile marker 608, was detected at least three times before the pipeline ruptured, in 2005, 2007 and 2009, according to documents Enbridge filed with PHMSA over the years. But each time, Enbridge decided it wasn't significant enough to require repairs within 180 days.

Ten days before 6B ruptured, Enbridge applied to PHMSA for another extension. It asked for an additional two and a half years to decide whether 6B should be repaired or replaced.

On the same day Enbridge applied for that extension, Richard Adams, the company's vice president of U.S. operations, assured a congressional subcommittee on pipeline safety that Enbridge was well prepared for an emergency.

"Our response time from our control center can be almost instantaneous, and our large leaks are typically detected by our control center personnel," Adams told the lawmakers. "They can view that there is a change in the operating system, and there are provisions that, if there is uncertainty, they have to shut down within a period of time, and that would include the closing of automatic valves."

The emergency response plan the company keeps on file with PHMSA is more specific. It says a rupture on the Lakehead system would be detected within five minutes and the damaged segment closed in three minutes.

The real-world test of Enbridge's emergency plan began at 5:57 p.m. Eastern Daylight Time on July 25, 2010, about the time the LaForge family was eating Sunday dinner. In a control room 1,500 miles away in Edmonton, Alberta, Enbridge was stopping the pumps on 6B as part of a scheduled, 10-hour shutdown. The company was waiting for more oil to fill storage tanks at the start of 6B in Griffith, Ind., so a full shipment could accumulate before pumping resumed.

One minute later, a high-priority alarm sounded in the control room, indicating that pressure had dropped to zero near Marshall. Another alarm triggered Enbridge's safety system and automatically halted the pumps at Marshall. Over the next five minutes, three more high-priority alarms signaled pressure problems on the line. Then a sixth alarm sounded, signaling a discrepancy between the volume of oil entering and exiting the pipeline.

At first, the control room operators weren't particularly concerned, according to a control room timeline and other documents recently released by the National Transportation Safety Board, or NTSB. They thought a large bubble had formed between batches of crude, a problem that often resolves itself. They figured the bubble would last until they restarted the pipeline early Monday morning.

The operators were so confident of their diagnosis that when their 12-hour shift ended at 8 p.m. they didn't mention the six alarms to their replacements, according to the NTSB documents.

Back in Calhoun County, however, noxious odors were beginning to permeate the summer night. At 9:25 p.m. local residents started dialing 911. One complained about a "very, very, very strong odor, either natural gas or maybe crude oil." Another described a house "asphyxiated with the gas smell" and asked if it was safe. Firefighters and local utilities checked the area for gas leaks, but found nothing. A Michigan Gas technician reported that he smelled petroleum.

The 911 calls continued as Sunday slipped into Monday.

At 4 a.m., controllers in Edmonton restarted the pipeline as scheduled and pumped oil up the line with the force of a firehose. Over the next hour, six more alarms went off.

At approximately 5 a.m., they shut 6B down again.

A pipeline analyst in the control room said 6B should be started again with more pressure, so oil would fill the line and overcome the bubble they thought was triggering the alarms.

"I guess there's two choices here, either consider it a leak or try it again?" the control center operations supervisor said.

"Just call it a false alarm," the analyst said.

As they prepared to restart the line for a second time, one of 6B's operators said he thought there was a leak. But others disagreed, and at 7:10 a.m., 6B's pumps kicked into gear again.

Four more high-priority alarms sounded as pumping continued for at least 45 minutes.

At 7:48 Monday morning—about the time Lorraine LaForge was telephoning John to tell him the smell near their house was even worse—the shift leader called for almost doubling pressure on 6B. But the extra power wasn't available, so they shut down the line again.

Two of the control room operators agreed they had never experienced a situation quite like this.

"Whatever, we're going home and will be off for a few days," one of them said. They left the control room a few minutes later.

The next shift took over, aiming to restart 6B as soon as extra power was available so they could clear what they still thought was a bubble from the line. At 9:49 a.m., they heard some reassuring news from Marshall: The Enbridge electrician who inspected the pump station and general vicinity hadn't detected any leaks or unusual odors.

The leak wasn't discovered until 11:17 a.m., when an employee for a Michigan utility company called Enbridge's emergency number with the bad news. Oil was pouring into Talmadge Creek, about three-quarters of a mile from the pump station, he said.

At 11:45 a.m., an Enbridge employee arrived at the site and confirmed the leak.

InsideClimate News asked Enbridge to answer a question that the NTSB timeline raises: If the company didn't know about the leak until 11:17 a.m., why had its workers gone to the LaForge residence at 10 a.m. and tested the air in the family's home?

A company spokesman said he couldn't answer that question, or any other question about the chronology of events, while the NTSB's investigation is ongoing.

Pipeline operators are required to report spills to the National Response Center in Washington, D.C. "at the earliest practical moment" following "discovery of a release." This notification is considered crucial to any cleanup response because the NRC alerts state and federal agencies to unfolding disasters.

Enbridge first tried to contact the NRC just after 1 p.m., according to the NTSB documents. The company had already alerted its own public affairs office in Houston about the spill 15 minutes earlier.

Because the NRC line was busy, Enbridge didn't get through until 1:33 p.m.—almost two hours after it had confirmed the spill and more than three hours after Enbridge workers urged the LaForges to leave their home. The company reported a spill of 819,000 gallons of oil.

Three minutes after Enbridge finished talking with the NRC, the center had contacted 16 agencies.

By this time, the same oily muck that had darkened the LaForges carefully tended lawn was sloshing over the banks of Talmadge Creek and coating tree trunks, flowers and soil along the Kalamazoo River. Jay Wesley, a fisheries specialist with the Michigan Department of Natural Resources, was already on the scene, trudging along the floodplain and collecting oil-coated muskrats and turtles in cardboard boxes and plastic bins.

Everything reeked of petroleum. Residents were on edge.

Deb Miller was driving home from her event-planning job in Battle Creek that evening when she saw several hundred people clustered on 12 Mile Road bridge. The bridge across the Kalamazoo River is in the village of Ceresco, about five river miles west of John LaForge's home. It offers a dead-on view of the Ceresco Dam, a local landmark.

Miller and her husband, Ken, had raised their two daughters in Ken's childhood home, which sits just 300 feet below the dam. They had built a deck off the back of their nearby flooring and carpeting business so they could enjoy watching fish swimming just under the river's surface.

The crowd parted so Miller could inch her car across the bridge and turn into her driveway. An overpowering odor of boiling hot asphalt assaulted her nostrils before she even opened the car door.

Miller joined the spectators on the bridge. Together, they watched an alarming brown mist rise as river water the shade of a dark chocolate malt tumbled 13 feet over the dam.

"We knew instantly by the smell and the color of the river that something had happened," Miller said, wrinkling her nose at the memory. "And whatever it was, it was huge."

Enbridge rushed workers to the creek as soon as the spill was confirmed. But even as they positioned absorbent boom on the water's surface and dug culverts to divert the oil, they suspected they wouldn't be able to stop it from surging into the river just a couple of miles away. Flooding from four days of heavy rain made the oil-soaked water almost impossible to contain.

The 175-mile Kalamazoo River is a treasured recreational area. After the federal Clean Water Act was passed in 1972, paper mills, wastewater treatment plants and other polluters had been forced to rein in their once-deadly discharges. Some stretches were so pristine that canoe paddlers could feel transported back to the 18th or 19th century. If rivers had personalities, Wesley, the fish expert, would have classified the pre-spill Kalamazoo as "natural and wild." In 2000, he and a team of scientists had documented it as home to 102 species of fish, 23 species of mussels and clams, 218 species of birds, 40 species of mammals and 40 types of amphibians and reptiles.

Keeping the oil out of this important resource was crucial. But the EPA, which was taking command of the cleanup, was also looking at the bigger picture.

The Kalamazoo is not a drinking water source. But about 115 river miles west of Marshall it empties into Lake Michigan. Together with the other four Great Lakes, Lake Michigan provides drinking water for at least 26 million Americans and close to 10 million Canadians. If the lake became contaminated, a local disaster would escalate into a regional catastrophe.

The EPA and Enbridge also worried about a stretch of the river near the city of Kalamazoo, about 43 river miles west of Marshall.

Polychlorinated biphenyls, better known as PCBs, were embedded in the river where a factory had dumped them years ago. The area had been declared a Superfund site, and nobody was sure what might happen if oil mixed with PCBs, which are known carcinogens.

The cleanup teams had two advantages as they planned their strategy. The break had occurred just minutes from Enbridge's maintenance facility in Marshall, so some cleanup equipment was immediately available. Marshall is also close to Interstates 94 and 69, so more apparatus could be trucked in quickly from Battle Creek, Kalamazoo, Lansing, Detroit and Chicago.

Dozens of federal, state and local officials converged at a makeshift command center in an Enbridge building near the center of town.

Durk Dunham, Calhoun County's emergency management services director, was confident this would be a quick in-and-out operation. He figured vacuum trucks would quickly remove the oil and everybody would be home for dinner that night.

But when Dunham surveyed the devastation from a helicopter later Monday—and saw pure black instead of a ribbon of river—he realized his initial assessment was wrong. His eyes teared up when he saw the extent of the devastation.

"It was heartbreaking," he said. "There wasn't much being said on that helicopter."

By the time Jim Rutherford, Calhoun County's public health officer, arrived that afternoon from his office in Battle Creek, the oil had overwhelmed the creek. Despite the best efforts of the cleanup crews, it was surging into the Kalamazoo River.

Rutherford, just two years into his job, was bewildered by what he saw. He and his staff were prepared to deal with tornadoes and other severe weather but they knew next to nothing about oil spills. Until that afternoon, Rutherford hadn't even known that an oil pipeline passed near Marshall.

"We were pressing Enbridge as to what their plans were," he said about those early chaotic hours at the command center. "They only had a middle manager there and he was like a deer in the headlights. Yes, EPA was there, but we really needed Enbridge to call the shots."

The officials had two questions to answer—fast.

Could a spark ignite a chemical explosion—a major concern at any oil spill? And did the vile-smelling air pose a health risk for nearby residents?

Answering the first question was relatively easy. Using monitors that measured the mixture of oxygen and hydrocarbons in the air, the EPA determined that the likelihood of an explosion was low to non-existent.

Finding a definitive answer to the second question was more daunting.

Every type of crude oil, including diluted bitumen, is made up of hundreds of chemicals, and many of them evaporate into the air after a spill. Scientists don't fully understand how some of these chemicals affect humans. During a congressional hearing on the spill, Scott Masten, a scientist at the National Institute of Environmental Health Sciences, would testify that "the potential for human health effects exist. However, understanding and quantifying these effects requires further study. There has been relatively little long-term research into the human health effects from oil spills."

One chemical commonly found in crude oil—benzene—is of particular concern, because it can cause health effects at low concentrations and over short periods of time. Studies have shown that people regularly exposed to benzene for several years can develop leukemia and other cancers.

The Natural Resources Defense Council and other environmental organizations have long contended that dilbit contains more benzene than conventional oil, but it's hard to know whether that's true. Little research has been done on dilbit, and most of that work was conducted by the industry and is considered proprietary information.

Workers with the EPA and Enbridge joined Michigan health officials in using an assortment of hand-held monitors to check the air for benzene, a standard procedure at any big oil spill. Some types of monitors, which they usually had access to, weren't available that first day because they were still at the BP oil spill.

The readouts in Marshall fluctuated dramatically. The monitors detected benzene levels that ranged from below 50 parts per billion (ppb) to as high as 200 ppb. Some alarming spikes—6,250 ppb and even 10,000 ppb—showed up over patches of oil on the water and away from homes.

Rutherford huddled with federal and state health experts to try to figure out what these numbers meant. Should they evacuate the hundreds of people who lived near the river?

As Calhoun County's health director, Rutherford was responsible for making that decision. But he felt overwhelmed. Until now, his primary focus had been coordinating food inspections and school nurse programs for the county's 136,000 residents. His health department didn't have access to monitoring equipment. In fact, only one of his 70 employees is dedicated to emergency preparedness.

"People need to understand that at a local level, we're totally dependent on state and federal resources in a situation like this," he said recently. "That's a reality."

For help, Rutherford turned to the federal and state health experts, people he would later describe as his "superheroes." But they couldn't provide any easy answers because no federal benzene guidelines applied specifically to their particular crisis.

The federal Agency for Toxic Substances and Disease Registry calculates that an average person can be exposed to 6 parts per billion of benzene—the rough equivalent of two tablespoons of liquid in an Olympic-size swimming pool—for up to a year without long-term health effects. The agency uses 9 ppb as the benchmark for up to two weeks exposure.

Another set of benzene guidelines, drawn up by a coalition of federal agencies is usually used for workers dealing with a short-term emergency. Those guidelines say that people can be exposed to up to 200,000 ppb for eight hours without increasing the risk of long-term health effects.

The health experts who gathered in Marshall weren't exactly sure how long the benzene would linger, but their expertise told them it would be longer than eight hours but shorter than two weeks—and definitely less than a year. So what were they to do?

People were already calling Rutherford's office, local hospitals and the Poison Control Center to complain about headaches, sore throats, nausea and vomiting—all symptoms that the Centers for Disease Control and Prevention has linked to benzene exposure. But other, less dangerous chemicals found in oil can also cause those symptoms.

For more accurate data they needed air sampling equipment, which requires more time to produce results but is more sophisticated than hand-held monitors. Mark Durno, an EPA deputy incident commander for the spill, said that the EPA team, veterans of many oil spills, considered this an ordinary spill and saw no need to rush sampling equipment to Marshall on Monday.

Late Monday Enbridge gave the EPA a Material Safety Data Sheet, or MSDS, a federally mandated document required for hazardous substances that are transported or used in a workplace. But the three-page MSDS didn't offer much guidance. Nowhere did it mention "bitumen" or "diluted bitumen" or "dilbit." The only clue that the oil in 6B might be different from conventional oil were references to "diluent" and "condensate," two terms that refer to the chemicals added to dilute the bitumen. But nobody seemed to recognize that the words indicated this was not ordinary crude.

Durno said the MSDS confirmed their assumption that 6B was carrying regular heavy crude oil. The EPA had supervised the cleanup of almost 8,400 spills since 1970, and the Enbridge supervisors at the scene did not hint that this spill might be different.

Rutherford and the other health care experts considered everything they had seen and learned that day. They agreed that an evacuation wasn't needed—at least not yet.

The monitoring was still continuing as Rutherford drove home in the wee hours of Tuesday morning. He pondered how long the oil would dirty the river of his childhood—and how far he would have to travel to outpace the hideous stink that soured the still summer air.

"It was kind of numbing, like being in a dream," he said. "Were we ever going to be able to get a handle on this?"

2. YOU CAN'T JUST EVACUATE AN ENTIRE COUNTY

On Tuesday, the day after the spill was detected, oil was still streaming from Talmadge Creek into the Kalamazoo River near Marshall.

Six inches of rain between Thursday and Sunday had turned the normally sedate river into a roiling brown torrent that overflowed its banks by several feet. The creek, usually only five or six feet wide and a foot deep, was at least 100 feet wide.

The EPA officials who had gathered in Marshall still thought they were dealing with the light crude oil that usually flows through U.S. pipelines. As veterans of other spills, they were certain they were prepared for this one. What they didn't know yet was that 6B, the pipeline that ruptured, had been carrying bitumen from Canada's tar sands region.

Enbridge's president and chief executive officer, Patrick Daniel, perpetuated the mistaken belief that this would be a routine cleanup. On Monday, Daniel had flown in from Enbridge's Calgary, Alberta, headquarters in the company jet. In an interview the next day, he said much of the oil could be sucked off the water's surface with vacuum trucks and that only a "minuscule" amount might sink below the surface.

"To tell you the truth, it's lighter than water so it sits on top of the water," he said.

Days of confusion followed the spill, with federal and state officials basing their cleanup decisions on the erroneous assumption that the oil was ordinary crude. It was an assumption that Enbridge did not correct. Federal regulations do not require pipeline operators to disclose the specific type of crude oil their lines carry. The nonprofit Pipeline Safety Trust and other organizations have urged the government to change that policy since Canadian dilbit was first pumped into the United States more than a decade ago.

Two deadlines the EPA set Tuesday reflected the agency's confidence in a quick turnaround. Enbridge was ordered to clean up the wetlands near the broken pipeline by Aug. 27. The creek, the river and all shorelines were expected to be oil-free by Sept. 27. Both of the orders mentioned only oil—not dilbit.

The agency's overarching objective was keeping the oil from reaching the spot where the Kalamazoo empties into Lake Michigan, an important drinking water source. The EPA was also concerned about the PCB Superfund site near the city of Kalamazoo.

While the scientists worried about protecting Lake Michigan from the oil, health experts fretted about the oil's effect on people living along the Kalamazoo's banks.

Benzene readings picked up by hand-held monitors were still swinging wildly. Readings ranged from less than 50 parts per billion, a level that didn't worry the health experts, to 3,000 ppb. The highest readings were in areas where oil was being recovered.

Jim Rutherford, Calhoun County's public health director, huddled with state and federal health experts. They had no idea how long the benzene would linger. And they still hadn't found any clear guidelines on whether people should be evacuated in these circumstances.

Finally they decided to create their own benchmark for evacuation, based on their analysis of the available scientific information.

The Michigan Occupational Safety and Health Administration lists 500 ppb as the workplace benzene limit. Using that standard, plus the federal standards they had studied earlier—and taking into account differences between workers and a general population that included children, the sick and the elderly—they decided on Wednesday to set 200 ppb as the benchmark for evacuation.

Rutherford would order an evacuation Thursday if monitors continued to show benzene readings of 200 ppb or above, they agreed. As the county health director it was also up to him to make the final call and to decide if it should be a mandatory or voluntary evacuation.

That Wednesday night before Rutherford headed home, the EPA reported that the size of the spill was at least one million gallons. That figure exceeded Enbridge's Monday estimate of 819,000 gallons.

Even after three days of working double shifts, sleep didn't come easily for Rutherford that night.

At home, he walked for miles under the stars, sorting out his burdens. Being thrust into the limelight as a local health official was scary. He knew some people thought he wasn't acting quickly enough. But they weren't in the war room grappling with a multitude of unknowns. Was it practical, or reasonable, to displace elderly people and families with young children when hotel rooms were already at a premium because of the enormous influx of cleanup workers? Plus, he and the other health professionals didn't have any hard scientific evidence that temporary exposure to 200 ppb of benzene did, indeed, pose a danger.

All along, his priority had been to protect people's health, not compound the chaos the spill was already causing.

"You can't just evacuate an entire county," he said recently, recalling those days of indecision. "It's easier said than done."

Rutherford walked until long past midnight, rehearsing how he would deliver the news if benzene levels were still high on Thursday.

The next morning he saw the test results that had come in during the night and knew what he had to do.

Although benzene levels were generally dropping, hand-held monitors still showed levels of 200 ppb or higher at some locations. Most of the readings had dropped below the level of concern, but there also were single measurements of 200 ppb, 250 ppb, 500 ppb and 1,350 ppb.

The more sensitive sampling equipment had arrived, and the first results would be ready later that day. But Rutherford decided not to wait. It was time to call a press conference and start evacuating people.

Residents of 61 riverside homes north and northwest of the rupture site were asked to leave because of "higher than acceptable levels of benzene." It was a voluntary, not mandatory order, because Rutherford didn't want to have to force people from their houses.

Rutherford also announced that people living within 200 feet of the river between Talmadge Creek and the Kalamazoo County line shouldn't use their well water for cooking or drinking. Tests showed no evidence of groundwater contamination, but he didn't want to take any chances. Enbridge agreed to provide bottled water.

Workers from the county and state health departments fanned out along the river to deliver the evacuation notices in person. If nobody was home, a notice was stuck on the front door.

Twenty-seven households in the evacuation zone refused to leave. But more than 100 families outside the zone moved out—some of them driven out by the stench before the evacuation was announced.

The evacuation notice offered hotel options, told them how to arrange accommodations for their pets, and encouraged them to save their receipts, so Enbridge could reimburse them. Enbridge also offered to help the uninsured with medical bills, whether they evacuated or not.

Rhonda Stepp, an administrative assistant at Marshall High School, learned her house was in the evacuation zone when her retired husband called her at work. She hurried home so they could gather a few belongings before heading to her parents' house in Battle Creek.

"When they tell you to pack up what can't be replaced, you're just thinking, 'Oh my God, what do I take?'" Stepp said. "I took pictures off the wall and the contents of our safe."

On the day the evacuation began, Enbridge gave EPA officials and other responders at the command center a second Material Safety Data Sheet. Like the first MSDS, it didn't mention dilbit. Again, Enbridge did not volunteer that information.

Mark Durno, the EPA deputy incident commander, said in a recent interview that if Enbridge had provided more specific information about the chemical makeup of the oil, the EPA would have rushed sampling equipment to the scene so sampling could have begun Monday.

Environmental organizations such as the Natural Resources Defense Council have long contended that dilbit contains more benzene than conventional oil. But it's difficult to determine if that's true.

The first air sampling data arrived from the lab that afternoon. It confirmed what the hand-held instruments had already indicated—although most of the benzene levels were below 50 ppb, some were as high as 550 ppb. Readings taken next to oil recovery sites ranged from 1,450 to 10,000 ppb.

The EPA was able to deliver one piece of positive news on Thursday. Although the oil had spread through more than two miles of Talmadge Creek and about 36 miles of the river, workers had managed to stop it before it reached the city of Kalamazoo. That meant that the PCBs buried in the river at the Superfund site wouldn't be disturbed—and that the drinking water so many people depended on from Lake Michigan was no longer at risk.

Susan Hedman, who directs EPA Region 5 in Chicago, was upbeat when she spoke with reporters on Sunday.

"I am happy to report significant improvement of the spill site, at the creek and the river," she said. "Oil continues to be removed and we have not seen any further contamination."

Enbridge had dispatched 730 workers to Marshall by the end of the first week. That didn't include the hundreds of local, state and federal experts still flocking to the scene. More than 69,000 feet of containment boom arrived, along with 43 boats, 48 oil skimmers, 79 vacuum trucks, 19 tanker trucks and 77 mobile storage tanks.

Helicopters zoomed overhead. Airboats plied the river. The grind of internal combustion engines added to the cacophony as transfer trucks hauled goo sucked out of the river.

Deb and Ken Miller's tiny neighborhood in Ceresco had been

transformed into a staging area complete with Dumpsters and a temporary dining area for workers. An ambulance and a fire truck were stationed near the bridge on 12 Mile Road, where buses and vans unloaded swarms of workers tasked with collecting oil. Decked out in white biohazard suits, they looked like space explorers. Sheriff's deputies set up a barricade at the bridge, and residents traveling that route risked arrest if they didn't stop.

Enbridge moved the command post to Walters Elementary School. Workers propped up their laptop computers on cardboard boxes and wedged themselves into chairs designed for grade-schoolers.

Daniel, the Enbridge CEO, apologized repeatedly for the damage his company had done in Calhoun County, where the rancid odors of oil were still powerful. At meetings, one-on-one talks and media interviews, he reassured residents that the company was committed "to cleaning up anything and everything" the oil had touched.

"We are responsible for the cleanup and we will be here until you are happy in this community… that we have completed our responsibilities," Daniel said.

Daniel promised John LaForge and several others who lived near the rupture point that Enbridge would "make them whole," by buying or building them new homes away from the river.

The Millers declined Enbridge's offer to move to a hotel. The closest hotels were already booked and they needed to watch over their store, their elderly dog and a homebound neighbor. To keep the stink at bay, they shut their windows and blasted the air conditioning.

Every time Deb Miller looked out her window, she fumed about the pain a broken oil pipe had inflicted on her community. She wondered why so few local authorities had known of 6B's existence and worried about the impact the oil was having on neighbors up and down the river.

She was still taking oral chemotherapy as part of her treatment for breast cancer in 2002 and she wondered if chemicals in the oil would compromise her health.

On Monday, Aug. 2, Miller and hundreds of other residents filed into the Marshall High School gymnasium for a public meeting organized by the EPA. Enbridge provided carpeting, so the folding chairs wouldn't scratch the gym floor. But company officials weren't invited to attend, because the EPA wanted to make sure, as it does after every such disaster, that the public understood that state and federal oversight agencies were in charge of the cleanup, not the company that had caused the problem.

Hedman, the EPA Region 5 director who opened the meeting, was still upbeat.

"We will continue working until your river looks like this again," she said, showing a PowerPoint image of the pre-spill Kalamazoo.

Most of the audience applauded enthusiastically. But Deb Miller wasn't in any mood to clap. To her, it seemed people were responding to wishful thinking rather than reality. She was dismayed that they weren't allowed to take the microphone and vent their concerns. Instead, they were directed to the cafeteria, where booths had been set up so they could speak privately with various officials.

In her one-on-one meeting, Miller told an EPA employee about a mass of oil that had accumulated in a river alcove near her carpet store. The official listened attentively and promised to send workers to investigate.

Still, Miller headed home that night feeling she had wasted her time. She wanted someone to spell out what kind of financial compensation would be available to those directly affected by the spill and she wanted to know when the oil would be cleaned up.

She also wanted assurances that the foul air wasn't jeopardizing people's health.

"We were given a spiel, then herded into areas to ask questions," Miller said. "We're not scientists. How do we know what to ask? That's what made so many of us resentful, like you cannot trust that our federal government is going to tell you everything. We don't know exactly what kind of oil is in the river and you have a gut feeling that they haven't been forthright."

The day after the meeting, Enbridge rolled out a program to buy properties along the polluted section of the river and creek. More than 310 properties, about half of them homes, were eventually eligible. Owners were given a year to accept or reject the offer.

By then, the terrible smell was abating. Experts at the National Oceanic and Atmospheric Administration later said most of the chemicals that had been added to dilute the bitumen probably evaporated by Aug. 4.

While Enbridge was reaching out to the community, it was also rushing to get 6B back on line. At least three U.S. refineries had been forced to reduce production, because they needed 6B's oil.

The company was losing money, too. Though Enbridge spokespeople didn't want to discuss it, the company's annual report states that earnings were down $85 million in the second half of 2010 for costs associated with the 6B spill.

Extricating the ruptured pipeline from the oil-saturated wetlands near John LaForge's home took more than a week. The two 20-foot pieces were trucked to the National Transportation Safety Board facility in Ashburn, Va., so they could be studied as part of the spill investigation. Enbridge pulled new pipe from its stock in Marshall and welded it into place.

On Aug. 9, two weeks after the spill occurred, Enbridge asked the Pipeline and Hazardous Materials Safety Administration (PHMSA) for permission to restart 6B.

PHMSA rejected the request less than 24 hours later.

The plan lacked "sufficient technical details ... to permit a conclusion that no immediate threats are present elsewhere on the line that require repair prior to any re-start of the pipeline," PHMSA said in its letter to Enbridge. The agency wouldn't approve any restart plan that "did not include excavating and exposing additional pipe and repairing or replacing additional pipe as necessary."

Among the flaws PHMSA listed was Enbridge's failure to "determine, investigate and remediate as necessary, at least four additional anomalies in Line 6B" that were similar to conditions near the spot where the Marshall leak occurred. Line 6B had several hundred corrosion defects and Enbridge had exercised its legal option to reduce pressure while it decided whether to repair or replace the line.

On Aug. 10, the Millers temporarily closed their carpet and flooring business. With the road in front of their store blocked off because of the cleanup, customers couldn't reach them.

By then, volunteers and workers were removing oil from 83 turtles, 66 Canada geese, 12 ducks, three swans and four muskrats at a vacant warehouse Enbridge had turned into a rescue center. They had already cleaned and released 22 turtles and a frog.

More than 99,000 feet of boom was now positioned in 37 spots between the creek and Morrow Lake. Another 250,000 feet of boom was ready—just in case.

On Aug. 17, Rutherford, the county public health officer, lifted the July 29 voluntary evacuation order because benzene readings were consistently below 6 parts per billion. He advised riverside residents to continue using bottled water for cooking and drinking.

At about the same time, cleanup crews began to notice something they hadn't seen at spills involving light crude oil.

The surface of the river was clearing in some places, a sign of progress. But when an EPA employee disturbed a clear patch of water near Morrow Lake—the dammed lake west of Marshall where they had finally stopped the oil—he noticed that tiny flakes of tar floated to the surface and formed a small oil sheen.

Closer to Battle Creek, crews with the Michigan Department of Natural Resources also noticed an odd phenomenon. When they disturbed the sediment at the river's bottom with their hip waders, globules of tarry oil popped up and created similar but larger sheens.

To determine whether these were isolated incidents or signs of a deeper problem, workers lowered absorbent material wrapped in chicken wire into the river to see what it captured. They also shook up the sediment with hand-held poles to see what floated to the top.

They were shocked by what they learned. Tar balls the size of marbles were being swept along the river's bottom with the clay, sand and other organic material that is normally caught up in river currents. Basically, the tar balls were bouncing downstream, stopping only when a deep pool, an eddy or a man-made barrier like a dam halted the ride. At low points in the riverbed, they were settling into as much as six inches of sediment.

Mark Durno, who has 20 years of experience with the EPA, had never seen anything like it.

"We had no idea sinking oil would be such a problem," Durno said. "Not only was this material submerged but it was mobile and moving along the river bottom."

At first, the scientists thought they could mount sonar or other high-tech instruments on boats or helicopters and map exactly where the oil had sunk. But the depth of the river, the type of sediment and the nature of the oil made that impossible.

Instead, teams of specialists had to resort to the laborious process of manually recording every square inch of the oiled river. Wielding hand-held poles, they poked the sediment to gauge how much oil they found. Each point was assigned a GPS (global positioning system) reading and added to a GIS (geographic information system) database. Using this digital map they could estimate the oil's footprint and volume. Over time, they could see where it moved and measure the effectiveness of their cleanup techniques.

This unusual twist in the cleanup operation was discussed at daily meetings attended by Enbridge and the government agencies supervising the cleanup. But Durno, who attended all the meetings, said Enbridge never volunteered the information that the oil was not light crude but Canadian dilbit.

What was happening at the spill site is now clear. After 6B ruptured, the liquid chemicals that had been added to dilute the bitumen began evaporating, and the heavy bitumen began sinking. When the surface of the river started clearing, it wasn't necessarily because the oil was gone, but because it had disappeared from sight.

The Natural Resources Defense Council, a powerful advocacy organization that opposes the Canadian tar sands industry as well as the Keystone XL pipeline, already suspected that 6B was carrying bitumen from Western Canada's tar sands fields.

Kari Lyderson, a former Washington Post reporter who was writing for the NRDC's monthly magazine, spoke with Enbridge's Daniel several times in August and asked if the oil in 6B was tar sands oil, or bitumen. She said he told her several times that it was not.

In a teleconference call with Lyderson and other reporters, Daniel implied that the oil in 6B wasn't tar sands oil because it had been extracted by steam distillation rather than mining. On that same call, however, he acknowledged that the oil was so thick that it had to be thinned by a third with light crude before it could be pumped through pipelines.

The NRDC attacked Daniel for "trying to be cute with his language."

A few days later, the CEO backpedaled on the tar sands issue. "What I indicated is that it was not what we have traditionally referred to as tar sands oil," he told the Michigan Messenger. "If it is part of the same geological formation, then I bow to that expert opinion. I'm not saying, 'No, it's not oil sands crude. It's just not traditionally defined as that and viewed as that.'"

As far as the EPA was concerned, the semantics of the debate didn't matter much. What did matter was the challenge the agency now faced. Bitumen lay at the bottom of a major U.S. river, a river that also happened to be at flood stage because of recent rains. The oil had to be removed. But how could they complete that cleanup mission without destroying the waterway they were trying to save?

Less than a month after the dilbit spill in Marshall, Enbridge's image took another knock. On Aug. 17 the Pipeline and Hazardous Materials Safety Administration fined the company $2.4 million for violating safety regulations on a pipeline in Clearbrook, Minn., which like 6B is part of the company's Lakehead system. It involved a November 2007 incident in which two company employees were killed after repairs caused leaking crude oil to ignite. PHMSA said, "Enbridge failed to safely and adequately perform maintenance and repair activities, clear the designated work area from possible sources of ignition, and hire properly trained and qualified workers."

The company received even more public scrutiny when the initial Aug. 27 deadline for cleaning up the Marshall spill came and went, unmet. By then, however, the EPA was beginning to understand why Enbridge was so far behind.

"It's safe to say we had a set of circumstances that combined to give us some challenges," said Ralph Dollhopf, who was leading the agency's efforts in Marshall. "At the onset of something like this, you rarely have details on the scope of work required. As Enbridge progressed, we learned how much oil was out there."

3. A MOVING TARGET ON THE RIVER BOTTOM

As the fall of 2010 approached, John LaForge could still smell tar when he drove by his old home on Talmadge Creek.

LaForge had lost hope that he and Lorraine would someday return to the house where they had lived for 28 years and raised four children. Tire tracks left by heavy equipment had scarred and muddied the lawn LaForge once tended so carefully.

The cleanup of North America's biggest dilbit pipeline spill was behind schedule and LaForge's property in southwestern Michigan, about a quarter mile from where an Enbridge pipeline had split open on July 25, was ground zero. More than 2,050 workers had flocked to Marshall. Parking was such a hassle at Kate's Diner, where he ate breakfast before work, that he worried regulars would stop patronizing the restaurant.

LaForge began negotiating with Enbridge for the company to buy his property. In September, he and Lorraine, along with their daughter and her three young children, left the two hotel rooms they'd shared for 61 days and rented a house while they looked for a place to buy. Enbridge footed the $12,000 hotel bill and agreed to pay their rent. All the moving was taking a toll on Lorraine. She was still recovering from the emergency gallbladder surgery she'd undergone while they were living in the hotel.

The LaForges salvaged photographs, dishes and hardwood furniture from their old home. But the oil stink had permeated their mattresses, clothing, books, toys, rugs and upholstered furniture. They left it all behind.

"How do you replace your granddaughter's little dress from her first day in kindergarten?" LaForge said, looking back on that difficult transition. "You put your sweat and heart into a place and then somebody comes along and destroys it. It's painful."

The spill was adding stress to Deb Miller's life, too.

She and her husband, Ken, finally re-opened their carpet and flooring store in October, two months after the spill forced them to shut it down. They had no intention of selling their house or business, even though both buildings were located near Ceresco Dam, another focal point of the cleanup. Enbridge offered to pay their rent if they temporarily relocated their business, but the offer didn't cover the cost of moving their inventory. The Millers said no. Instead, they accepted an "inconvenience" payment for lost income.

Watching the cleanup drag on was turning Miller into an activist. Her bout with breast cancer had alerted her to health issues, and she feared that the toxicity of the oil might have jeopardized residents and emergency responders in ways that scientists didn't understand. She filled a three-ring binder with 8-by-10 color photographs documenting the mess at the dam and carried it to meetings and strategy sessions with neighbors.

"First responders are our neighbors, our dads and our brothers," she said. "What training were they provided? Our local agencies were tasked with responsibilities they were in no way equipped to handle."

In mid-September, Miller took her photos to Washington, D.C., where she and five other Calhoun County residents testified before the House Transportation and Infrastructure Committee. The chairman, Jim Oberstar, was a Democrat from Minnesota, where another section of Enbridge's Lakehead pipeline system is located. Two representatives from Michigan served on the committee: Mark Schauer, a Democrat who represented the Marshall area, and Candice A. Miller (no relation to Deb Miller), a Republican from the eastern part of the state.

It was Deb Miller's first trip to the nation's capital. She was nervous, but determined to be heard. She labored almost three weeks on her 19 pages of testimony. Congressional staffers had told Miller and her neighbors to "write from the heart."

"I knew I had to do what I had to do," she said recently. "My message was that I'm not going away. We told our stories because somebody had to put a face on what the impact of this spill was."

The Sept. 15 hearing in the Rayburn House Office Building lasted seven hours. EPA Administrator Lisa Jackson was among the witnesses. So were National Transportation Safety Board Chairman Deborah Hersman and Enbridge CEO Patrick Daniel.

Miller sipped water to control the nagging cough she'd had since the spill.

"I was an innocent bystander," she said when it was her turn to sit behind a microphone and address the committee. "I did not choose to breathe that foul air. I did not choose to lose a summer to … vacuum trucks, fan boats, and helicopters and strangers on my riverbank, not to be able to utilize our pool in our back yard for lack of privacy. I did not choose to close my business, and I certainly did not choose to watch the geese struggle while covered in oil. Enbridge made that decision for me.

"I sincerely hope this spill will ensure that you (Enbridge) will be more responsible with the maintenance of all of your pipelines, even if it means replacing them all," she added. "I pray they will remain closed until that can be determined how safely to restart them."

Another Calhoun County resident, Michelle BarlondSmith, told the committee that when she and other residents of a Battle Creek trailer park sought health care for spill-related symptoms, an Enbridge representative told them they had to sign a waiver form. They later learned that the form gave the company access to their entire medical histories.

The trailer where BarlondSmith lived with her husband, Tracy, was just 200 to 300 feet from the oiled river, she told InsideClimate News. They spent several weeks at a hotel to escape the stench, which she said made her feel dizzy and sick to her stomach.

In a transcript of the testimony, Schauer, the representative from Battle Creek, asked BarlondSmith if she was comfortable with the company having access to her medical records.

Ms. BARLONDSMITH: To be very frank with you, one of the side effects that you have with this is you do not think clearly... I read over it twice very quickly. I gave it to my husband. He glanced at it because he was going to go to the doctor also.

Mr. SCHAUER: He is not an attorney, I take it, or a health care provider?

Ms. BARLONDSMITH: Unfortunately, he is not an attorney and I wish he was. But I signed it because I was told if you wanted to see the doctor, you must sign this.

Schauer grilled Daniel, the Enbridge CEO, about the medical release form. Schauer and Oberstar had sent a letter to Daniel on Sept. 1, demanding that Enbridge stop asking uninsured residents to sign the waiver.

Mr. SCHAUER. So have you stopped the use of this form?

Mr. DANIEL. I don't know that offhand. I can get back to you and confirm that.

Mr. SCHAUER. Well, and I also request—and I think I requested this in writing—that you rescind all of those that have been signed. Would you agree to do that?

Mr. DANIEL. Yes.

Mr. SCHAUER. Thank you.

Rep. Candice Miller pressed Daniel about a defect on a section of 6B in her district where the pipeline is buried under the St. Clair River, a vital drinking water source for northern Michigan.

The dent had been identified in August 2009 and was serious enough to meet PHMSA's criteria for repair within 60 days. But 11 months later, it still wasn't repaired. And Miller wanted to know why.

In his testimony, Daniel explained that because "the site is very difficult to access," Enbridge decided to lower the operating pressure while conducting "a comprehensive engineering assessment."

"The likelihood that that dent will cause a leak is very remote," he assured the committee. "It is smooth, without evidence of corrosion or cracking. The pipe at that point is twice as thick as normal and is protected by concrete and engineered gravel. Nonetheless, Enbridge is committed to replacing or repairing that segment of pipe, and we will submit our proposed plan to the regulator by the end of this month."

Daniel also reiterated the promise he had made so often since he arrived in Michigan the day the spill was detected.

"I am personally committed and our company is committed to doing everything that we can to make up to the people in Marshall and Battle Creek for the mess that we made," Daniel said. "We are working very diligently to meet the September 27th deadline for cleanup of the spill, in conjunction with the EPA and all of the coordinating agencies… You have my commitment that we will be there to make your constituents happy that we have done the right job."

When Deb Miller's plane landed in Michigan that night, she was almost giddy after watching how committee members held Enbridge accountable.

"I'm not naïve enough to think that everything would be resolved that day," she said. "But I walked out of there with a ray of hope that maybe somebody was listening."

But Miller was scared, too. She'd always assumed that the oil that was polluting the river in her back yard was ordinary crude. But in casual conversations away from the microphone that day, people had called it "diluted bitumen," a term she'd never heard before.

"Hearing the oil being described as a totally different product knocked my feet out from under me," Miller recalled. "My first reaction was to cry. Then I wondered, 'What else have they lied to us about?' To this day, that is why I am so frustrated with EPA and Enbridge. Nobody knocked on my door and told me I was in danger."

Miller wasn't alone with her fears.

By early September, local residents had dialed the hotline Enbridge set up the day of the spill at least 9,400 times. The hotline, as well as the county health office, local hospitals and the Poison Control Center, had been flooded with questions about what harm the stinky air might be causing.

A survey of four riverside communities that the Michigan Department of Community Health conducted within a month of the spill found that almost 60 percent of the 550 people interviewed experienced headaches, breathing difficulties, coughs, vomiting, anxiety or other health problems.

The federal Pipeline and Hazardous Materials Safety Administration allowed Enbridge to reopen pipeline 6B on Sept. 27, two months after the spill was detected. The agency limited it to carrying 10.2 million gallons per day instead of 11.3 million gallons per day. But Enbridge was back in business.

The cleanup wasn't proceeding as rapidly.

Enbridge missed its Sept. 27 EPA deadline—the one that required it to rid the creek, river and shorelines of all oil. A new deadline was set for Oct. 31.

Close to 30 miles of boom was now positioned along the river. But more oil kept turning up. It saturated soil and plants along the floodplain. It contaminated small islands along the river. It was embedded in up to six inches of underwater sediments.

"I truly believe the characteristics of this material is the reason we still have such a heavy operation out here," Mark Durno, the EPA deputy incident commander told Michigan Public Radio. "Because it was a very heavy crude, we ended up with a lot more submerged oil than we anticipated having to deal with....If you'd shovel down into the islands you'd see oil pool into the holes we'd dig."

Durno had become a fixture at Pastrami Joe's, a popular deli. Twelve- and 16-hour workdays meant he stayed in touch with his wife and two young children back in Ohio with text messages and brief phone calls. Every day, his wife e-mailed him photos of the home remodeling project they had begun about the time of the spill. The way the cleanup was proceeding, he figured he wouldn't be leaving Marshall any time soon.

Federal regulations require culpable parties—in this case Enbridge—to restore waterways to their pre-spill state. But how was the company going to remove every bit of submerged oil from 36 miles of the river when it hadn't even been able to thoroughly clean more than two miles of Talmadge Creek?

Tracking and removing the transient blobs of bitumen that had sunk to the bottom of the river was especially frustrating.

In October, the EPA directed Enbridge to experiment with dredging a three-acre area above the Ceresco Dam, which was inundated with oil, because it was so close to the rupture. Crews operating excavators dug for about three weeks and carted away 5,500 cubic yards of oil-soaked sediment, enough to fill 27 semi-trailers. They also removed, decontaminated and then returned 14 million gallons of water to the river.

They managed to extract the bulk of the oil. However, that brutal but efficient operation wasn't an option elsewhere on the oiled river. All of that gouging would destroy fish habitat and ruin underwater beds where mussels feed and breed.

Other traditional cleanup methods were also proving harmful.

Ripping out oil-coated islands and oil-ravaged logs and plants deprived fish of vital shelter. And the steady beat of waves caused by so many boats on the water eroded the banks where muskrats and beavers burrowed for shelter.

Gradually, everybody agreed that they had to treat the river as a living organism, not as an entity to be conquered.

Enbridge began developing more gentle techniques. Workers on foot, in boats or in marsh buggies used rakes with metal tines, rototiller blades, chain drags or air- and water-spraying wands to gently agitate the oil by hand. Then they vacuumed it up or collected it with nets, booms and absorbent pads.

On average, the Kalamazoo is only about three feet deep, so instead of always using boats with standard engines that could tear up the shallow river, Enbridge brought in flat-bottomed "airboats" powered by raised aircraft-type propellers and engines.

Progress was slow. Nobody was surprised when Enbridge failed to meet the EPA's Oct. 31 deadline for removing all submerged oil from the river. A few days later, the company increased its estimate of how much oil had spilled from 6B from 819,000 gallons to 843,444 gallons.

But the news wasn't all grim.

On Nov. 5, Jim Rutherford, Calhoun County's public health officer, announced that people who lived near the river could once again drink and cook with their well water. No pollutants had been found, although the testing would continue.

Despite this reassurance, Deb Miller stuck with the bottled water. Instead of cooking her family's Thanksgiving dinner in Ceresco, she moved the celebration to her younger daughter's house a few miles away.

By this time Enbridge also had also managed to skim, vacuum and sop up most of the visible oil in the creek and river. It was a small step in the right direction, even though everybody was sure oil remained hidden in the creek and its banks.

"It's kind of like doing an initial surgery," Ralph Dollhopf, the EPA incident commander, said about the first pass at cleaning the creek. "It's done to get the gross amount of oil and get the situation stabilized. We knew residual oil would be identified afterward and we'd have to come back to meet long-term requirements."

The cleanup ramped down for the winter. By mid-December, only about 200 workers were on-site.

Before the year ended, Enbridge announced that it had been able to recycle 766,288 gallons of oil recovered from the spill site. Instead of sending it to a landfill, the company was able to return it to the pipeline terminal in Griffith, Ind., where it was again pumped through 6B.

In the spring of 2011, teams of scientists continued the tedious process of mapping the submerged oil. The digital snapshot that emerged confirmed their fears. Tar balls the size of marbles were still piling up in low spots on the river bottom.

Roughly 200 acres, an area about the size of 150 football fields, were still tainted with oil.

Three landmarks were identified as "oil magnets." One was above Ceresco Dam, next to the Millers' business, where they had dredged in October. The second was near the dammed Mill Pond in Battle Creek. The third was at the delta of Morrow Lake, where the river flows into a dammed recreation area before it reaches the city of Kalamazoo.

"The submerged oil is a real story, it's a real eye-opener," the EPA's Mark Durno told the Natural Resources Defense Council's OnEarth magazine. "In larger spills we've dealt with before, we haven't seen nearly this footprint of submerged oil, if we've seen any at all."

They were back to the problem they had started with: How would they tackle submerged oil that was a moving target?

John Sobojinski, the engineer who had supervised Enbridge's operation in Marshall since November 2010, said beating the river to death didn't make sense.

"You would have to run bulldozers and excavators down 38 miles of river and take out everything to get every last bit of oil," Sobojinski said. "The river would never recover."

Enbridge and the EPA devised a new plan that Dollhopf described as the "locate, clean up and repeat" approach. Instead of trying to scour the entire river bottom, they would let the tar balls roll into the three spots the scientists had pinpointed as oil magnets. As the tar balls accumulated, they'd go in and extract them. It was frustrating to have to wait out the oil, but the evolving science supported their patience.

"At a minimum, we're writing a chapter in the oil spill cleanup book on how to identify submerged oil," was how Dollhopf described the challenges they faced. "We're writing chapters on how it behaves once it does spill (and) how to recover it."

In some areas Enbridge continued using the gentle cleanup techniques it had developed in the fall of 2010 to capture underwater oil. But elsewhere crews also tried a more mechanized—and harsher—approach to agitate and collect the dilbit. They fitted excavator buckets with rototiller blades, pulled chain drags and air- and water-spraying wands and rototiller blades behind boats, and equipped pontoon boats with excavators that could pull chain drags.

In June 2011, the EPA gave Enbridge a new deadline: Finish the river cleanup by Aug. 31. But the company missed that deadline, too. More than 800 workers remained on the job.

The EPA's Susan Hedman no longer sounded so optimistic. "Capturing and cleaning up this heavy oil is a unique challenge," she told reporters a year after the cleanup began. "No one at the EPA can remember dealing with this much submerged oil in a river."

As the cleanup slogged on, the people of Marshall were growing accustomed to the presence of the workers and the economic benefits they brought to their little community, which bills itself as the "City of Hospitality." Hotels were often full and workers were spending money at the Stagecoach Inn, the Dark Horse Brewery and Schuler's Restaurant and Pub, a historic landmark downtown.

Enbridge—whose slogan is "Where Energy Meets People"—tried to solidify that feeling of goodwill by donating money to an assortment of causes.

The company upgraded a park near Battle Creek that had been closed by the spill and paid for a new bridge between the park and a large river island. It built fishing and boating piers at five other recreational sites and set up an endowment fund to maintain them. It donated $100,000 to the Calhoun County Trailway Alliance's hiking trail project and promised another $100,000 if the alliance raised matching funds.

Other gifts included $45,000 to United Way branches in Marshall, Battle Creek and Kalamazoo; $50,000 to the Marshall school district and $20,000 to the county fairgrounds.

Most people appreciated Enbridge's efforts. The vice president of a local conservation club, which received about $25,000 from the company, said that despite the tragedy of the oil spill he thought Enbridge was "really attuned to the environment."

A woman who accepted what she described as Enbridge's "generous offer" to buy her home on Talmadge Creek said "Everything they did was a class act. Everything."

Others, including Deb Miller, viewed Enbridge's generosity as a public relations gimmick.

"People say, 'Well, Enbridge is trying its best,'" Miller said while standing on the porch of her carpet business overlooking the river. "Well, maybe its best isn't good enough. There's no end in sight. What's going to happen 10 years from now if the oil is still in the river?"

In October 2011, Enbridge CEO Patrick Daniel was named "Canada's Outstanding CEO of the Year." The president and CEO of Caldwell Partners, the law firm that founded the award, described Enbridge as "an exceptional community supporter having invested in hundreds of charitable and non-profit organizations across Canada and the United States."

On an unseasonably warm day in November, John LaForge drove with a an InsideClimate News reporter past his old house near Talmadge Creek. Using money from his settlement with Enbridge, he had built a new house—as well as pole barns for his excavating and garbage-hauling businesses—four miles away. The family had moved in over the summer.

LaForge said he felt his settlement with Enbridge had been a fair one. But he still winced when he saw the cracks crisscrossing his once-immaculate concrete driveway—and when he noticed that someone had cut down the crabapple tree he and his wife had planted in memory of their son, Justin, who had died in a car accident at the age of eight.

From the house he drove to a nearby neighborhood of ranches, colonials and luxury homes that had been built along the Kalamazoo over the last several decades. He had excavated some of the basements.

As he navigated the long horseshoe-shaped road, he periodically pointed to empty houses that Enbridge now owned.

"People make jokes that we live in Enbridgeville because they've bought everything," LaForge said. "They don't realize what people went through. That company thinks money can buy anything."

With another winter approaching, the Enbridge workforce tapered off to about 450. The focus would be on meeting EPA's new deadline for the creek cleanup: March 31, 2012.

The only way to be sure the creek would be oil-free was to strip away the contaminated stretch, a little more than two miles. Essentially, crews would be building a new creek from scratch. Dredging—the technique they'd considered too severe for the river—made sense here.

Contractors pieced together mazes of corduroy roadways and navigated their excavators, front-end loaders, graders and dump trucks along the floodplain. Then they scraped the oily creek bed and its banks down to the bone, scooping out 21,578 cubic yards of dirt. Finally, they hauled in tons of "new" dirt, shaping it to follow the path the creek had traveled before the spill.

When they finished, the water ran as clear as ever through the reinvented portion of the creek. The only hint that something traumatic had occurred were the yards of landscaping cloth and erosion control blankets spread out to protect the newly planted native grasses and other vegetation. Tiny trees planted during the mild winter were already sprouting new roots.

Enbridge met the March deadline. And Jay Wesley, the fish expert with the Michigan Department of Natural Resources who has spent 16 years studying the watershed, was confident the creek would bounce back.

The only question was how long it would take Mother Nature to right herself. In 2000, a survey found 11 species of fish and 192 individual fish in that segment of the creek. A few weeks after the spill, Wesley and other natural resource specialists had counted just three species of fish and 53 individual fish.

On April 18, 2012—21 months after 6B ruptured—the first mile of the tainted section of the Kalamazoo River was opened to the public for boating and swimming. Jim Rutherford, the Calhoun County's health director who had called for a voluntary evacuation after the spill, was also responsible for making this decision.

To celebrate their accomplishment, the EPA's Durno and a dozen other federal, state and local officials climbed into kayaks and paddled the cleaned-up portion between Perrin Dam and Saylor's Landing.

Durno, who takes a dip in Lake Erie every New Year's Day when he's home in Cleveland, wore his wetsuit. He slipped out of his kayak and swam for a few minutes in the 60-degree water.

Durno said he didn't see or smell any oil during his swim. But neither he nor anyone else involved in the cleanup suggests that all the oil is gone.

Technically, restoring the river to pre-spill conditions would mean removing every last tar ball, no matter the cost. But scientists have realized for months that would be foolhardy.

"Do we sterilize the river and destroy its ecology to restore it?" asked Durno "That's the key question."

Teams of specialists using poles are now doing another survey, their third, to determine how much oil remains in the river's bottom. The results won't be in for a few weeks, but the EPA's Dollhopf said they are "definitely seeing significant reductions from last year and the year before."

To plan their next steps, the scientists and the cleanup experts have sliced the river into ecological sections according to patterns of oil contamination, types of wetlands, and species of animals, plants and trees. Each section will be cleaned with the technology that works best for its unique situation. Heavily oiled sections might be tackled with more intrusive methods. Lightly oiled areas may be treated with nothing more than some bundles of pine and fir trees placed underwater to trap the tar balls that are still bouncing along the river's bottom.

"Some of those scenarios may involve leaving oil behind, so it's unlikely that every last drop of oil will be removed," said Dollhopf, who still works out of the cluster of temporary trailers near the rupture site, where the cleanup command post has been housed since the fall of 2010. "We don't want to cross over the balance point of the benefits of oil removal and the harm of oil recovery. We always have to weigh that."

On June 21, Rutherford opened about 34 more miles of the river. The only section that's still closed is a small stretch at the delta of the Morrow Lake, which is marked off with buoys. The EPA estimates that 1,148,229 gallons of oil have been recovered so far. Enbridge still maintains that its ruptured pipeline released only 843,444 gallons.

Rutherford said the water in the open section of the river now meets all necessary health and safety requirements. A study by the Michigan Department of Community Health said people who come into contact with the oil might suffer some skin irritation, but they won't experience long-term health problems.

Information kiosks at ramps along the river are now stocked with brochures citing that study and telling people what to do if they see or touch oil. The kiosks also have disposable wipes for removing oil from skin or boats.

After three and a half years of deliberating about whether to repair or replace 6B, Enbridge recently asked the Michigan Public Service Commission for permission to replace the line and almost double its capacity. Replacing 6B through Michigan and Indiana will cost close to $1.9 billion. It will be 36 inches rather than 30 inches wide in most places and capable of pumping up to 21 million gallons of oil per day.

The expansion is needed, Enbridge says, to meet the growing demand of U.S. refineries for cheap Canadian dilbit.

InsideClimate News intern Kathryn Doyle contributed to this report.

EPILOGUE: CLEANUP, CONSEQUENCES AND LIVES CHANGED IN THE DILBIT DISASTER

How big was the spill? Was it tar sands oil? Who will pay? How did animals and ecosystems fare? What happened to the people most affected? What fines could Enbridge face for its oil spill in Marshall, Mich.?

The size of any fines will depend, in part, on how much oil was spilled when Pipeline 6B ruptured.

The EPA's latest estimate, released on June 7, is that 1,148,229 gallons (or 27,339 barrels) have been recovered since the cleanup began on July 26, 2010.

Enbridge maintains that it spilled only 843,444 gallons (20,082 barrels), an estimate the company hasn't changed since November 2010.

The discrepancy between these numbers matters, because penalties levied under the Clean Water Act are figured on a per-barrel basis.

Enbridge's civil penalties could reach $4,300 per barrel of oil spilled if the government can prove gross negligence under the Clean Water Act. If gross negligence can't be proved, civil penalties could still be as high as $1,100 per barrel. Criminal penalties under the Clean Water Act could be up to twice the losses associated with the spill.

Defining those losses is a gray area because no case law exists, said David Uhlmann, a professor at the University of Michigan Law School and former chief of the U.S. Department of Justice's environmental crimes section. Losses incurred by victims of the spill and the cost of the cleanup are likely to be counted, but lost revenue to Enbridge would not.

Generally, the government chooses either criminal or civil penalties except in the most egregious cases, such as the Gulf spill, Uhlmann said.

Fines also could be levied against Enbridge under the Pipeline Safety Act. If federal authorities find that the company violated any of the standards set by that legislation, it could face civil penalties of $200,000 per violation per day. At a minimum, the penalties would likely include the days 6B was leaking.

Criminal penalties under the Pipeline Safety Act are similar to those under the Clean Water Act: up to twice the losses associated with the spill. Federal authorities would have to prove knowing and willful violations to levy a criminal fine under the Pipeline Safety Act but only would need to show simple negligence under the Clean Water Act.

The results of the National Transportation Safety Board's investigation of the spill will likely factor significantly into the levying of penalties. That report will be released on July 10, according to the NTSB.

Just as the government isn't required to accept Enbridge's estimate of the number of barrels spilled, neither is Enbridge required to agree to the government's estimate. If the federal government and Enbridge go to court, the number of barrels spilled would be determined at trial.

Uhlmann predicts a settlement, not a trial, is the most likely outcome with this case. If so, the number of barrels spilled and the fine levied per barrel are negotiable. Uhlmann expects the government would insist on its estimate in exchange for a slightly lower fine per barrel.

Enbridge's price tag for the spill is already $765 million, with $650 million covered by the company's insurance. That insurance doesn't cover fines or penalties.

What happened to the oil and debris that was hauled away from the spill site?

Enbridge said it was able to recycle 766,288 gallons (18,245 barrels) of the oil it vacuumed up shortly after the spill. It returned the oil to 6B, which reopened two months after the spill.

Contaminated soil and debris that was collected was disposed of off-site. According to the EPA, so far crews have disposed of 17,109,012 million gallons of oiled water and 187,041 cubic yards of oil-contaminated soil, downed logs and other debris.

The 766,288 gallons of oil that Enbridge recycled—plus the amount of oil the EPA estimates was trapped in the oily water and debris—is how the agency calculated its latest figure of 1.14 recovered gallons.

Was the Oil in 6B Tar Sands Oil?

About 75 percent of the oil that spilled from 6B was a type of dilbit called Cold Lake blend. In an email to InsideClimate News, Enbridge spokesman Jason Manshum made the same argument that CEO Patrick Daniel made shortly after the spill when he spoke with Natural Resources Defense Council reporter Kari Lyderson: That the oil in 6B shouldn't be classified as oil sands crude, which is also known as tar sands crude.

"Oil sands and heavy crude such as Cold Lake are typically blended at the production site with diluent to assist in the transport of the crude," Manshum said. "Cold Lake heavy crude is not located in the oil sands region around Fort McMurray, Alberta and retrieving this crude is done by way of drilling not by open pit mining. It is part of overall formation, but since it's not from the same area, it [is] not typically associated with oil sands crude."

Both the industry and the spill investigation documents say otherwise. The Cold Lake and Fort McMurray regions of Alberta contain separate oil sands deposits that are part of the same general formation. These oil sands are mined for the bitumen, which is classified as either "extra heavy" or "heavy" crude oil under the American Petroleum Institute gravity scale. Once the bitumen is diluted with liquid chemicals, the resulting dilbit is considered heavy crude oil.

Imperial Oil, one of the largest petroleum producers in Canada, places Cold Lake crude under the "oil sands" section of its Web site. The company's website states that "Cold Lake bitumen is located more than 400 metres below the surface out of the ground and we extract it by injecting steam into the oil sands to thin the heavy bitumen and enable it to flow to the surface through wellbores. Cold Lake produced more than 160,000 barrels of bitumen a day in 2011."

The steam injection method described by Imperial Oil is a drilling technique used to extract bitumen from oil sands deposits. The other commonly used technique is open pit mining.

Crudemonitor.ca, an industry website on crude oil chemistry supported by the Canadian Association of Petroleum Producers, lists the Cold Lake blend under the "Heavy sour-Dilbit" category. The website says that "Cold Lake production is bitumen based and requires the use of steam to release the bitumen from the underground reservoirs, and the use of diluents to meet pipeline viscosity and density specifications."

Documents from the National Transportation Safety Board's investigation of the 6B spill also refer to bitumen and dilbit. A report prepared for Enbridge by AECOM, a technical consulting firm, describes the Cold Lake blend as "a heavy crude of bitumen and blended with diluents."

In August 2010, less than two weeks after the spill, scientists at the National Oceanic and Atmospheric Administration wrote that the Cold Lake oil "is a blend of bitumen (API 11) and condensate (API 69). The blend has an API of about 21 and is made up of 70% bitumen and 30% condensate [diluents]."

Is Enbridge expanding other pipelines besides 6B?

In mid-April, Enbridge filed paperwork with the Michigan Public Service Commission, seeking its approval to replace 6B, which is part of the company's Lakehead system. The commission's decision is expected early next year, and the company hopes to complete the project by autumn 2013.

Part of the new 6B will be 30 inches in diameter, just as the old pipeline is, but the bulk of it will be 36 inches wide. The new line will be capable of pumping up to 21 million gallons of oil per day—almost double 6B's pre-spill daily capacity of 11.3 million gallons. The price tag to replace 6B in Michigan and Indiana is nearly $1.9 billion.

Enbridge has also proposed other pipeline projects, including new lines to transport Canadian oil to U.S. refineries. Some will carry tar sands oil and others will carry light crude from the Bakken oil fields in North Dakota, Montana and southern Saskatchewan.

Eastern Canadian Refinery Access Initiative.

Announced in mid-May, this project would reverse the flow of Enbridge's 240,000 barrel-a-day Line 9 between Montreal, Quebec to Sarnia, Ontario. Tar sands crude and Bakken light crude oil would move east from Sarnia to Montreal, and possibly eventually to Portland, Maine, via a separate pipeline. Part of the reversal could be completed by next spring.

The project also includes expanding the company's 500,000 barrel-a-day Line 5 from Superior, Mich. to Sarnia by roughly 50,000 barrels a day, as well as an expansion of 80,000 barrels a day on Enbridge's Toledo Pipeline, Line 17, which runs from Stockbridge, Mich. to refineries in Detroit and Toledo, Ohio. Both of those lines are part of the Lakehead system.

Seaway Crude Pipeline System Project.

Enbridge and Enterprise Products Partners completed the reversal of the Seaway pipeline last month, so it can move crude from Cushing, Okla., to Texas refineries. By mid-2014, the companies plan to twin the pipeline and to add 450,000 barrels per day of capacity to the Seaway system, which would nearly double its current capacity.

Flanagan South Pipeline Project.

Expected to be online by mid-2014, the proposed Flanagan line would run 600 miles from Flanagan, Ill., southeast of Chicago, to Cushing, Okla. From Cushing, crude would then move to Houston and Port Arthur, Texas, on the Seaway pipeline system. Initial capacity would be 585,000 barrels a day.

How Did Land and Water Creatures Fare?

A year after 6B ruptured, data gathered by the U.S. Fish and Wildlife Service showed that, remarkably, a majority of the animals rescued and cared for by professionals and volunteers survived.

Nearly 98 percent of the 3,651 reptiles collected and cleaned were released to non-oiled sections of the river. The rate for birds was close to 75 percent, with 144 of 196 surviving. Mammals didn't fare as well. Only 23 of the 63 collected—or 36.5 percent pulled through. The dead included muskrats, raccoons, voles, skunks and at least one mink and one mole.

Jay Wesley, the fish specialist with the Michigan Department of Natural Resources, figured any fish trapped in the oil in July and August 2010 were pumped out of the contaminated area by cleanup crews along with the water and oil. While biologists noticed a few dead fish here and there in the river, Wesley hypothesized that most of the other fish out-swam the initial oil onslaught.

Wesley is optimistic that most of the creatures that count on Talmadge Creek and the Kalamazoo River are resilient enough to adapt to post-spill changes. But he continues to pay very close attention to fish reproduction and diet.

Some studies indicate that petroleum is detrimental to fish eggs. Scientists will have to study several generations of fish to figure out if the oil harmed fish eggs in the creek and river, he said, and that will take years to compile. Biologists are also monitoring whether fish and other creatures higher on the aquatic food chain have enough to eat. That requires tracking the plethora of minuscule aquatic bugs known as macroinvertebrates. Will the midges, mayflies, damselflies and stoneflies that typically crawl around on underwater vegetation or lurk in sediments at the bottom of creeks and rivers make a strong comeback?

What's Happening with the Cleanup?

Although most of the river is now open to the public, that doesn't mean the cleanup is over.

About 230 workers in boats and boots spread out along the river in mid-May, poking the sediment with poles to gauge how much oil is left. They expect to complete the assessment within the next week or so. Cleanup scientists have learned that it makes more sense not to begin mapping the underwater tar balls until the water is 60 degrees or higher, because that's when the oil is more mobile and apt to rise to the surface.

Ralph Dollhopf, the EPA incident commander supervising the cleanup, is still in Marshall full time, overseeing the work. Mark Durno, the agency's deputy incident commander, still makes occasional visits to the site.

Dollhopf can't yet say how many more months or years the agency will remain at the site. Once the EPA signs off, the Michigan Department of Environmental Quality will continue to monitor the creek and river for years, or even decades. That effort will include testing water and sediments. Enbridge will continue to foot the bill.

Update: Deb and Ken Miller

Business at the carpet store Deb and Ken Miller own in the village of Ceresco is about 30 percent below what it was before 6B ruptured. At the peak of the cleanup, they closed the store for two months because roads were blocked and so much heavy equipment clogged their neighborhood that their customers couldn't reach them.

In addition to her duties at the store and her job as an event planner in nearby Battle Creek, Deb Miller is also flourishing as an activist. She travels across the country to training sessions and forums as part of a new advocacy and outreach group formed by the nonprofit Pipeline Safety Trust, which is based in Bellingham, Wash. Miller laughingly refers to herself as "the commoner" among the cavalcade of engineers and scientists who were invited to join the group.

"I don't have the scientific knowledge, but not one of those scientists can talk about the pain I've been through these last two years," she said in a recent interview. "We need to put a face behind the pain. Otherwise, there will never be any accountability.

"It's not like I hate Enbridge. I'm just asking for safe transport of a product we all need. After what we've endured, I've learned that we need pipeline safety and accountability. That needs to go all the way to the top, so I'm hoping our state, federal and local agencies have learned something."

Miller wrote a letter opposing the Keystone XL pipeline when the State Department asked for public comment. If approved, that project would cross the nation's largest drinking water aquifer while carrying dilbit from Alberta to the U.S. Gulf Coast. She told federal authorities that the tragedy that unfolded in her community shouldn't be repeated along the Keystone XL route or anywhere else.

Miller also continues to press for funding for a long-term health study of residents and responders in Marshall. She's leery of a report released earlier this month by the Michigan Department of Community Health, concluding that contact with the chemicals in the submerged oil would not cause long-term health effects for humans.

The toxicology study was the final version of a report released in draft form in August 2011.

It found that "contact with the submerged oil will not cause long-term health effects or a higher than normal risk of cancer. At the same time, contact with the submerged oil may cause temporary effects, such as skin irritation."

Researchers reached these conclusions using a hypothetical worst-case scenario—a person immersing in the most oil-coated section of the river every day during the six-month recreation season.

The study accounted for children, the elderly and those with weakened immune systems. But Miller is concerned that it suggested people with pre-existing medical conditions consult with their doctors because potential effects of exposure on specific individuals are too difficult to predict.

"As a cancer patient, I can take preventive measures," Miller said. "But I couldn't do anything about this spill happening."

Update: John LaForge

John LaForge no longer wakes up to the babble of Talmadge Creek. Since moving into his new house in July 2011, his one-way drive to downtown Marshall has jumped from one-and-a-half miles to five miles.

Relocating hasn't snuffed out LaForge's entrepreneurial zest. Although his excavating, trash hauling and lawn mowing businesses are still the backbone of his ventures, he has branched out into raising chickens and beef cattle, and training helicopter pilots.

Watching oil despoil a landscape he has cherished for almost six decades motivated LaForge to explore alternative energy options. Memories of the water-pumping windmill on his grandparents' farm prompted him to invest in a Colorado-based company that promotes wind power.

He also plans to convert his new house into a demonstration project so fellow Michiganders can learn the ABCs of installing turbines and harnessing energy from the sun and wind instead of relying on fossil fuels trapped below the Earth's surface.

"Any fuel you have to burn, you have a contamination risk," LaForge said in a recent interview. "But when you see those turbines turning in the air, how is that going to hurt you? ... How much money do we want to keep spending on all of this crude oil? Let's get serious about wind and solar."

Update: Jim Rutherford

Jim Rutherford, the Calhoun County public health officer, said the pipeline disaster didn't spur local, state or federal officials to boost funding for his department—he said he'd need a magic wand for that to happen.

"If somebody would say to me, 'Would you be ready for an incident like this tomorrow?' Even after what we've been through this time, I would say certainly not," he said.

Local departments faced with these types of calamities still must rely almost entirely on the EPA and state agencies, he said.

"I'm older, balder and grayer because of this," Rutherford said. "We're still reeling from this incident almost two years later. But the reality is that tomorrow we wouldn't do things much differently than what we did today."

What galls him is the education he has received about how poorly regulated the pipeline industry is, he said. He's astounded that Enbridge is finally spending almost $1.9 billion to replace a faulty pipeline that was allowed to limp along for years—but only after the company spent at least $765 million on a cleanup.

He wonders how many other pipelines pose a similar risk to other communities.

"We're not even taking care of the antiquated infrastructure we have in the ground now," Rutherford said with a sigh. "Whenever I hear about a new pipeline being proposed anywhere, I shudder at the idea of it."

Stacy Feldman contributed to this report.

Follow-Up Reports

EPA WORRIES DILBIT STILL A THREAT TO KALAMAZOO RIVER, MORE THAN 2 YEARS AFTER SPILL

Enbridge needs to dredge accumulating oil from 100 acres of the river's bottom, EPA says. The work could take up to a year and cost tens of millions more.

By David Hasemyer

The hidden, long-term effects of the 2010 pipeline accident that spilled more than a million gallons of heavy Canadian crude oil into Michigan's Kalamazoo River became public last week when the EPA revealed that large amounts of oil are still accumulating in three areas of the river.

The problem is so serious that the EPA is asking Enbridge Inc., the Canadian pipeline operator, to dredge approximately 100 acres of the river. During the original cleanup effort, dredging was limited to just 25 acres because the EPA wanted to avoid destroying the river's natural ecology. The additional work could take up to a year and add tens of millions of dollars to a cleanup that has already cost Enbridge $809 million.

The EPA notified Enbridge of its proposed order on Oct. 3, saying the additional clean-up is "critical" and the work "should be conducted in an expeditious manner" to remove the oil before it recontaminates the river.

"The increased accumulation demonstrates that submerged oil is mobile and migrating, evidencing that submerged oil removal is warranted to prevent downstream migration ... ," Ralph Dollhopf, the EPA's on-scene coordinator and Incident Commander, said in the letter notifying Enbridge of the agency's findings.

In June an InsideClimate News investigation revealed that the cleanup of the Kalamazoo has been unusually difficult, because the pipeline that ruptured was carrying dilbit, a mixture of heavy Canadian bitumen that has been diluted with liquid chemicals, some of them toxic. Bitumen, also known as tar sands oil, has the consistency of peanut butter and is too heavy to flow through pipelines without being thinned with chemicals. When Pipeline 6B split open, the chemicals began evaporating and the reconstituted bitumen began sinking to the river's bottom.

"More than two years after the spill of diluted bitumen, this proposed order demonstrates that EPA is still tackling the problem of how to remove the heavy oil from the Kalamazoo River," said Sara Gosman, an adjunct professor of environmental law and policy at the University of Michigan Law School.

The EPA's determination that more cleanup is needed was based on the findings of a year-long survey of nearly 6,000 locations along the 40 miles of river contaminated when pipeline 6B ruptured in July 2010. Enbridge has until next week to request a conference with the EPA to discuss the additional work and 30 days to submit written comments.

Steve Hamilton, a Michigan State University professor who was among the experts who worked on the study, said the recommendation for dredging was driven by concern that during flooding the pools of oil could break loose and recontaminate parts of the river that have already been cleaned—or flow downriver into areas that were never touched by the gooey oil.

"We will never get all of the oil out [of the river]. It's impossible," Hamilton said. "The challenge is to determine when do you get to a point of diminishing returns where the eradication is too environmentally destructive to warrant the removal."

A spokesman for the EPA said the agency would not have any comment beyond the information contained in its proposed order and the letter it sent to Enbridge.

The EPA acknowledged in the proposed order that Enbridge had conducted substantial cleanup since the pipeline ruptured, but "despite these response actions, oil remains in the Kalamazoo River."

Enbridge did not respond to requests for comment for this story. But in an Aug. 24 letter to the EPA, the company said it did not believe that more dredging—especially in the area near the Ceresco Dam—was necessary.

"Enbridge's position is that we have reached a point of diminishing returns where further invasive activities would do more harm than good," Richard Adams, Enbridge's vice president of field operation in the United States, said in the letter.

"In fact, we strongly believe that such action solely for the purpose of aesthetics would both negatively impact the riverine environment and create a significant disturbance and inconvenience to local landowners and other river users."

The company also disputed the EPA's concern that oil is still pooling in the river, especially near the Ceresco Dam.

"[T]he most significant evidence of submerged oil has been sheen which, when collected, has amounted to a volume of less than 1 gallon of product in total during 2012," Adams wrote, referring to the area around the dam.

Deb Miller, who lives near the dam in the community of Ceresco said she sees rainbow sheens of oil floating on the surface when she walks along the river near the carpet store she and her husband own. Recently she ran a garden rake along the river's bottom and said that marble-sized globs of oil popped to the surface, accompanied by the sour whiff of petroleum.

"It's insane how much oil is still here," said Miller, who has testified before Congress about the spill's impact on her life.

Dilbit: The Unknown Factor

The National Transportation Safety Board blasted Enbridge in July for a "complete breakdown of safety" in the 2010 disaster, which is considered the largest inland oil pipeline spill in U.S. history. The report criticized the company for failing to make repairs despite knowing of the defects five years before the rupture. The Department of Transportation also imposed a record $3.7 million civil penalty. Enbridge paid the fine last month.

Enbridge has proposed replacing the entire 210-mile length of 6B from Indiana to Ontario, Canada, at a cost of $1.3 billion. But the project has faced resistance from landowners who are fighting the company's efforts to condemn their land and from lawsuits claiming Enbridge hasn't complied with all state and local regulations and environmental laws.

The study of the contaminated 40-mile section of the Kalamazoo that resulted in the EPA's directive began in 2011 and ended in August.

The EPA enlisted 14 federal, state and local organizations—including the U.S. Fish and Wildlife Service, the U.S. Geological Survey, and the Michigan Department of Environmental Quality—to perform the study as part of a Net Environmental Benefit Analysis to ensure the ongoing cleanup was sufficient and further ecological damage from the spill would be minimized.

Hamilton, the Michigan State University professor of ecology and environment, joined the team as a representative of the Kalamazoo River Watershed Council. He has done extensive research on the river and its flood plain and spoke to InsideClimate News not as a representative of the EPA but as one of the individual scientists who worked on the investigation.

Hamilton said the study relied on a technique called poling, where a long pole is used to churn up the bottom of the river to see if oil or residue floats to the surface. He said the poling identified about two dozen sections of the river where enough oil remained to be of concern. With those areas in mind, the scientists used a model of the river to simulate floods equal to the high water marks of the last 100 years, five years and the highest flood mark since the spill.

They were particularly attentive to the hundred year flood levels despite the statistical improbability of such a flood occurring.

"With climate change it might be more possible than the record might indicate," Hamilton said.

The recommendation for dredging was based on factors beyond aesthetics, Hamilton said. One of the scientists' primary worries was that not much is known about dilbit.

"This kind of crude oil is a complex mix of hundreds of compounds—some known to be toxic—that has not been studied much," he said. "We just don't understand the consequences well enough."

Congress has ordered a study, which is being conducted by the National Academy of Sciences, to determine whether dilbit is more likely than conventional oil to corrode pipelines. The study isn't expected to be finished until the summer of 2013.

Three Areas at Risk

The investigators decided that "sheen management"—a technique that uses booms to contain oil floating to the surface—was appropriate for most of the sections where they found pools of oil. But they concluded that dredging was the only solution for three areas of the river between Marshall and Kalamazoo, Mich. The vulnerable areas are upstream of Ceresco Dam, upstream of the Battle Creek Dam in the Mill Ponds area, and in the delta upstream of Morrow Lake. Together, they cover about 100 acres, an area about the size of 75 football fields.

Near the Ceresco Dam, the investigators discovered the area of submerged oil had increased from 20 acres to 23.5 aces and that oil globules were floating to the surface, according to the EPA's proposed order.

Because that area was subjected to what the EPA called "highly effective" dredging in 2010, the agency concluded that additional dredging would prove successful. The earlier dredging project lasted about three weeks and crews carted away 5,500 cubic yards of oil-soaked sediment from the river bottom, enough to fill 27 semi-trailers. An estimated 14 million gallons of water was decontaminated and returned to the river.

Mill Pond, the second section of the river cited for intense cleanup, presented more of a quandary for the EPA. Some sections shouldn't be dredged, the agency decided, because the digging and scraping would do too much damage to the sensitive ecology and because the submerged oil wasn't likely to move down river.

At the third proposed cleanup site, the Delta just upstream from Morrow Lake, the investigators discovered a "substantial expansion" of the submerged oil, with the plume now covering most of the two-mile length of the delta, an area of about 55.5 acres.

Hamilton said the scientists decided dredging was needed, because floods might dislodge the submerged oil and allow it to flow into a part of the Kalamazoo River unblemished by the spill.

"It would be wise to get at it now when it's practical before it either becomes lodged in small backwater areas or migrates into areas where oil has not been previously discovered," he said.

ANGRY MICHIGAN RESIDENTS FIGHT UNEVEN BATTLE AGAINST PIPELINE PROJECT ON THEIR LAND

A 2010 oil pipeline spill contaminated Michigan's Kalamazoo River. Now the line is being replaced, raising the ire of landowners along the route.

By David Hasemyer

The notice that arrived at Debbie and David Hense's home last September didn't seem especially alarming. Enbridge Inc. was going to replace Line 6B, the oil pipeline that leaked more than a million gallons of heavy crude into Michigan's Kalamazoo River in 2010. Since 6B runs through the Henses' 22-acre property near Fenton, Mich., some of the construction would be done there.

What the Henses didn't know, however, was that Enbridge intended to take an additional swath of their land for the pipeline—and there was little they or any of the other landowners who lived along the 210-mile route could do to stop it.

In addition to the existing 60-foot easement Enbridge already has through the Henses' property, the company wants another 25 feet—about the width of a two-lane highway—for the new pipeline. It also wants a temporary 60-foot easement for a work area.

For the Henses, this means the loss of a century-old stand of trees. In Oceola Township, Beth Duman will lose part of her back deck. In the town of Howell, Peter Baldwin will lose a section of the nature preserve he has nurtured for decades.

Today the Henses and other angry residents have become unlikely activists, determined to at least have a voice in the $1.3 billion replacement project.

In July a lawsuit was filed in a Michigan appeals court on behalf of property owners along the first phase of the project, where work has already begun. Five landowners filed the lawsuit anonymously because they fear retaliation in their negotiations with Enbridge. They're asking that approval for the project be revoked because, among other things, the notice they received from the Michigan Public Service Commission didn't tell them that Enbridge wanted more of their land.

More problems are brewing along phase two. Michigan's Public Service Commission had been expected to approve that segment in August. But the commission agreed to delay its decision until early next year, to give landowners more time to prepare for the hearing. One of the landowners, Jeff Axt, has founded a nonprofit called Protect Our Land And Rights (POLAR) to help with legal expenses. Opposition also is growing along the 60 miles of 6B that pass through Indiana.

Enbridge, Canada's largest transporter of crude oil, says replacing 6B with a parallel pipeline is "absolutely critical to the Michigan and U.S. refining industry." The line opened in 1969 and is part of Enbridge's Lakehead system, which delivers heavy Canadian crude oil to the United States and Ontario, Canada. Because the diameter of the new 6B will be bigger, Enbridge will be able to nearly double the amount of oil it can transport to U.S. refineries.

A recent investigation by InsideClimate News into the 2010 spill found that in the years leading up to the accident, federal regulators repeatedly cited Enbridge for corrosion problems on 6B. Enbridge spokesman Larry Springer said the new pipeline will be thicker, will be fitted with "enhanced" leak detection systems and will have computer assisted programs that constantly monitor the line.

"Enbridge uses thoroughly tested steel pipe that meets or exceeds all applicable standards and regulatory guidelines for quality and safety," Springer said in emailed responses to questions about the replacement project.

Springer downplayed the opposition the project faces.

"While there has been recent publicity and activity by special interest groups, most who live and work along the pipeline are not opposed to Enbridge's plans to replace Line 6B," he said. "While the media may choose to focus on controversial situations, Enbridge's actions show that we deal openly and honestly with all stakeholders, including landowners and local governments."

Springer declined to say how many property owners have not reached an agreement with the company or how many condemnation cases Enbridge has filed.

Resistance to the project has been so great that at one point Enbridge hired guards armed with semi-automatic pistols to stand watch near the backyards of recalcitrant farmers. The Alberta, Canada-based company also briefly contracted with the Livingston County Sheriff's Department to use off-duty deputies for security patrols, a tactic one local official called "a form of intimidation."

Jeffrey Insko, an American literature professor at Oakland University in Rochester, Mich., is so outraged about the company's "heavy-handedness" that he started the Line 6B Citizens' Blog. The pipeline crosses his two-acre property near Pontiac, Mich.

Some of Insko's anger is directed at public officials who he said have done little to protect the state's residents.

"We find it more than a little perplexing that Michigan—the state in which the most expensive inland oil spill in U.S. history occurred just two years ago—would be suffering from such a dearth of leadership," Insko said. "One would think, to the contrary, that politicians would be falling all over themselves to get tough with Enbridge, particularly after the release of the blistering NTSB report."

The National Transportation Safety Board report Insko referred to was released in July. It blasted Enbridge for a "complete breakdown of safety." Federal regulators imposed a record $3.7 million civil penalty on the company, which Enbridge paid yesterday.

Cleaning up the Michigan spill was especially difficult, InsideClimate News reported, because 6B carries dilbit, a mixture of heavy Canadian bitumen diluted with liquid chemicals, some of them toxic. When the pipeline split open near the town of Marshall, the bitumen began sinking to the bottom of the Kalamazoo River. Enbridge is still removing oil from the river more than two years later in a cleanup effort that so far totals $809 million.

One Family Fights Back

Insko has followed news of the spill and the cleanup with a mixture of anger and frustration. Still, like most landowners along the route, he grudgingly settled with Enbridge, rather than go through condemnation hearings he believed would be futile.

Enbridge's opening offer is usually $6,500 per acre according to attorney Kim Savage, who represents a number of landowners. The company negotiates separately for the land it needs for temporary workspace and for damages during construction.

No structures can be built and no trees planted on land Enbridge takes for its easement. Seasonal crops like hay are allowed, but if the company needs access to the land the crops can be removed and the owner paid market value.

The 60-foot work area reverts to the landowner when the project is complete. But during construction, Enbridge can clear it of all trees, crops or buildings. The company has said it will reseed workspace areas.

The Henses are among those who have rejected Enbridge's offers. Yet the company's workers showed up last month at the family's 22-acre property in Tryone Township, near Fenton, Mich. Debbie Hense, a chemist, got a panicked call at work from her 13-year-old son.

"Mom. Mom! They're cutting the trees!" he told her.

When Hense got home, a logging machine was on the disputed land, tearing away at a line of hickory, oak and cherry trees that she treasured. Some were 50 feet tall and 100 years old. She could hear the whine of blades cutting into the trunks.

"It was a terrible sound," she said. "I watched as the trees shook, cracked and then slowly fell over."

Hense got a restraining order that same day, temporarily barring Enbridge from working in the disputed area. That afternoon she positioned a folding chair in front of her remaining trees, just in case the workers came back. Before long, the 44-year-old mother of three was joined by about a dozen other people armed with their own lawn chairs. One came from Battle Creek, 100 miles away.

Two days later, Enbridge notified the Henses that it had begun condemnation proceedings for their land. The company also asked an appeals court judge to order the Henses to immediately allow tractors back onto the land, arguing that the couple was needlessly delaying the 6B project and jeopardizing oil deliveries to Midwestern and Canadian refineries. "The halting of the Project ... will result in irreparable harm not only to Enbridge, but more importantly to the public at large because the replacement of Line 6B has been determined to be and is in the public's interest," Enbridge said in the complaint it filed.

The judge declined Enbridge's request and set a Sept. 25 court date to hear the Henses' request for a permanent injunction. Enbridge then asked the judge to require the Henses to post a $612,500 bond, so the company could recoup any loses created by the delay. The judge will address that request at the hearing.

The Henses' attorney, Chris Christenson, said Enbridge's conduct is vindictive.

"They want to make an example out of my client and send a message to any others who have not yet reached agreements that if they resist in any way they face the same treatment," Christenson said.

Enbridge spokesman Jason Manshum declined to talk about the company's dealings with the Henses, saying the company doesn't discuss individual cases.

"The Greater Good?"

Most of the people who are fighting the 6B project know they are unlikely to get it stopped.

"Without pipelines the energy supply complex would collapse," said Bruce Bullock, director of the Maguire Energy Institute at Southern Methodist University's Cox School of Business in Dallas. "How would goods get to market? What would happen to transportation? Commerce is dependent on gasoline."

Enbridge made that point in documents it filed with the Michigan Public Service Commission.

"With no other high-capacity crude transportation system available to the region to connect to the same sources of production, it is important that Enbridge maintains its Line 6B to ensure reliable, safe and economical crude oil delivery both to Michigan's refinery and to refineries in nearby states and provinces that, in turn, produce gasoline and other petroleum-based products for Michigan consumers and businesses," the company said.

While pipelines may be inevitable, the landowners say their concerns deserve respectful consideration.

"What really bothers me is that we are forced to give up our land, our peace of mind for what is supposed to be the greater good," said Laurie Lentz, who lives on a 14.5-acre farm near Howell, Mich. "The greater good? We are just being used. I have a problem with that. It's big business as usual."

The new 6B will be buried under a hay field about 200 feet from the house where Lentz and her family have lived for 22 years.

Safety is her biggest concern. "In light of the catastrophe in Marshall, people are not anxious to have this toxic, nasty stuff running through their backyards under high pressure," she said.

Like many residents, Connie and Tom Watson, who live on 6.5 acres in Howell, Mich., are upset about the notice they got from the Public Service Commission.

Tom Watson, a 64-year-old Vietnam veteran, was coping with a serious medical problem—his transplanted kidney was failing—when the notice arrived at their house in September 2011. It said Enbridge would be replacing 6B and told them they could comment at a public hearing that month or file an electronic petition. But nowhere did it state that Enbridge would be taking more of their land.

An accompanying letter from Enbridge briefly mentioned land acquisition, in a single sentence near the end: "Enbridge will meet with various landowners and nearby residents to discuss acquisition of new right-of-way."

"The notice they received was so inadequate that an average person would not know what was beneath the surface," said Gary Field, one of attorneys representing landowners.

Field thinks regulators and politicians aren't raising hard questions about 6B's replacement because they are simply buying Enbridge's justification for the project.

"They are telling people and state regulators 'Hey wouldn't you rather have this nice, shiny new pipeline in your backyard instead of that old corroded one that could blow at any time?'" Field said. "No community wants to say no to that. And no politician wants to take that chance."

Only eight people showed up at the public hearing to protest the 6B project, and the Public Service Commission quickly approved it.

The Watsons thought no more about the pipeline until last spring when work crews began driving three-foot tall stakes into the ground to mark off a wide strip of their land for the new route.

"Had that letter been clear, you bet I would have made my feelings heard—loud and clear," said Connie Watson, who has refused Enbridge's offers to settle and is facing condemnation of her property.

In court papers filed in the lawsuit, Enbridge noted that the notice was written by the Public Service Commission and that any claim that it was inadequate is "outlandish."

Gary Kitts, the commission's executive director, said the agency's lawyers drafted the notice and followed language used in hundreds of other notices.

"It met the statutory requirements of the state for notification," he said.

Kitts said he couldn't comment on the landowners' complaints because he hadn't read the notice. He declined InsideClimate News' request that he review the two-page document. He said he was unaware of the court challenge and had no knowledge that landowners were outraged over Enbridge's right-of-way demands.

"I am certainly not going to go back and reassess a notice based on what may or may not be going on in the courts," Kitts said.

Negotiation or Condemnation

Kim Savage, one of lawyers representing the landowners, said about 10 percent immediately take what the company offers for their land. Others negotiate for months, with Enbridge sometimes upping its offer from $6,500 an acre to $8,000 an acre for farmland and $15,000 an acre for residential property. Some landowners have squeezed out payments up to $45,000 an acre, Savage said. Exact figures are hard to obtain, because landowners must sign non-disclosure agreements when they settle.

Beth Duman said the land agent who showed up at her front gate was blunt and arrogant. "Basically he said, 'If you don't cooperate we'll have to condemn your property.'"

Assuming that they would lose their land either way, she and her husband decided to bargain. They asked for $100,000 for the hundreds of trees and the strip of land they would lose on their 10 acres.

The negotiations went on for months.

"Every day it was on our minds," Duman said. "We couldn't get away from it. What are they going to do? When are they coming back? What do they want? It was awful living with that."

Because of the confidentiality agreement, Duman couldn't reveal what they finally settled for, "but I can tell you we got diddly squat."

Carol Brimhall, who lives on a 38-acre farm in Stockbridge, 15 miles from the Dumans, rejected Enbridge's $53,000 offer for her land and the 112 trees it had tagged for cutting. She was already angry with Enbridge, because she said the company had acted arrogantly a decade earlier when it took a strip of her property for a natural gas line. That time she settled. This time she decided to force the company to condemn her land.

"I told them, 'You guys don't get it. It's not about the money. I want my trees saved and my little animals protected,'" she said. "They couldn't buy me off."

On Aug. 1, Brimhall became the first landowner on phase one of the project to face condemnation proceedings. To win, she would essentially have to convince a judge that Enbridge didn't need her property. That's an especially high hurdle, given that the Public Service Commission has already decided that the project "is reasonable and in the public interest, and should be approved."

The hearing didn't take long, five minutes by her reckoning.

Her attorney argued Enbridge didn't need such a large swath of land – 85 feet wide.

Enbridge argued that it had the Public Service Commission's blessing to take as much land as it needed.

"The judge said 'OK. You want it. You got it.' and we were out of there," Brimhall said.

Brimhall fears Enbridge will be vindictive because she fought so hard. To save some of the trees on the land the company will take for workspace, she offered the use of a nearby natural clearings. But she said Enbridge rejected the offer, saying it was going to go ahead and cut down her trees.

Now Brimhall faces another day in court. This time she'll try to persuade the judge that her land is worth more than Enbridge is offering. She's asking for $125,000. But she knows the odds of getting that much are slim.

Enbridge wouldn't comment on Brimhall's case. Springer, the Enbridge spokesman, said the company has worked with many landowners over the last year and "reached mutually acceptable agreements as to compensation and location of the replacement segments."

"As with any major pipeline project, we sometimes cannot reach agreement with landowners or their legal representatives," Springer said. "Nevertheless, we have a long and demonstrated history of working diligently and fairly to resolve the vast majority of issues to the satisfaction of affected stakeholders."

A Township Stands Up to Enbridge

About six miles of 6B run through Brandon Township, a community of 15,000 that prides itself on the beauty of its gently rolling countryside. The headwaters of two rivers form in Brandon Township, including the Flint River, a major source of drinking water for the nearby city of Flint.

When the replacement project began, Brandon Township Supervisor Kathy Thurman and the township's trustees began hearing complaints from residents about Enbridge's plans.

The trustees invited company officials to a town hall meeting in early July, hoping to get details about the project and what residents could expect. But the answers weren't specific enough to reassure the local officials, and Thurman said they felt uneasy when they left the meeting.

"The health, welfare and safety of our people is the number one priority," Thurman said. "If we were to have a spill in Brandon it would be disastrous to our residents and others because of the pollution that would result.

"We got the feeling there was more to it than we knew. It was time we took a stand."

The township's attorney did some research and determined that because Enbridge would be using local roads, the township had legal authority under Michigan's constitution to require Enbridge to get local consent.

But when the lawyer told Enbridge what the township was proposing, the company made it clear that it didn't need permission from Brandon Township or any other local government, Thurman said.

The township decided not to push the issue.

"We just do not have the financial resources to be backed into a court battle with a company that has so much money," Thurman said.

Enbridge has an estimated worth of $30 billion. The township has an annual budget of $1.8 million.

Still, the trustees decided they had to do something.

In a unanimous vote on Aug. 20, they approved a resolution that asked Enbridge to meet six safety demands. It described the heavy bitumen that 6B carries as "extremely hazardous to the environment and all forms of life."

Enbridge responded with a letter outlining the pipeline's safety features and emphasizing that safety is regulated at the federal level.

The letter did not respond to two other requests in the township's resolution: That Enbridge guarantee that the existing 6B pipeline won't be reopened after the new one is in place, and that the company compensate the township for using its roads.

Hope Tempered by Pessimism

Insko, the literature professor and 6B blogger, sees Brandon Township's resolution as a small sign of hope in the landowners' fight to have their concerns addressed.

"Actions like Brandon's—action based on principle, regardless of whether they're overmatched– have the potential to initiate a groundswell of similar actions," he said. "If five, six, seven townships speak up, adding their voices to the public chorus, then our state elected officials might wake up—and at that point, Enbridge will have to begin to take notice."

But Insko is not always so optimistic. In one recent blog entry, he wrote, "It was as if Enbridge's sudden appearance in our township was just a change in the weather: a natural occurrence like rain or fog, something hardly worth mentioning, much less something anyone could do anything to change."

FEW OIL PIPELINE SPILLS DETECTED BY MUCH-TOUTED TECHNOLOGY

InsideClimate News analysis of a decade of federal data shows general public detected far more spills than leak detection technology.

By Lisa Song

For years, TransCanada, the Canadian company that wants to build the Keystone XL pipeline, has assured the project's opponents that the line will be equipped with sensors that can quickly detect oil spills.

In recent newspaper ads in Nebraska, for instance, TransCanada promised that the pipeline will be "monitored through a state-of-the-art oil control center 24 hours a day, 365 days a year. 21,000 sensors along the pipeline route relay information via satellite to the control center every five seconds."

Other companies make similar claims about their remote sensing technology, sometimes promising they can detect and isolate large spills within minutes.

But an InsideClimate News examination of 10 years of federal data shows that leak detection systems do not provide as much protection as the public has been led to believe.

Between 2002 and July 2012, remote sensors detected only 5 percent of the nation's pipeline spills, according to data from the Pipeline and Hazardous Materials Safety Administration (PHMSA).

The general public reported 22 percent of the spills during that period. Pipeline company employees at the scenes of accidents reported 62 percent.

Anthony Swift, an attorney who has spent years researching pipeline safety for the Natural Resources Defense Council, was taken aback by the findings. Swift's organization opposes the Keystone XL, and he said he had always known that leak detection systems didn't catch most of the spills. But "the fact that 19 out of 20 leaks aren't caught is surprising, and certainly runs counter to a lot of rhetoric we hear from the industry," he said.

Industry experts, however, were not surprised. Pipeline specialists interviewed by InsideClimate News said the findings are consistent with what they have observed.

"The reality of the science" is that there are limits to remote leak detection. "That's just the way it is," said Richard Kuprewicz, president of Accufacts, Inc., a consulting firm that provides pipeline expertise for government agencies, the industry and other parties. Kuprewicz has worked with TransCanada in the past, but is not involved with the Keystone XL.

Operators can feel pressured to "tell people things they shouldn't tell them because it's not true," Kuprewicz said. While the companies "may not be saying that with the intent of lying, the reality is, it's just real difficult to detect [releases] remotely."

TransCanada spokesman Grady Semmens answered questions about the Keystone XL in a series of emails. He said the pipeline's leak detection system will have "greater sensitivity" than is required by law. If the company can't identify the cause of a problem within 10 minutes, Semmens said the pipeline will be shut down and the affected section isolated "to immediately stop the flow of oil."

Leak detection is becoming increasingly important, because the industry plans to build thousands of miles of new pipelines over the next five years. Many of the pipelines will cross aquifers and rivers that are critically important for drinking water. Some of the projects, including the Keystone XL, will carry Canadian diluted bitumen, or dilbit, a mixture of heavy tar sands bitumen and light liquid chemicals. A recent InsideClimate News report on a 2010 dilbit spill in Michigan's Kalamazoo River revealed that the dilbit was much harder to clean up than conventional oil, because it gradually sank to the river's bottom.

The Michigan spill also showed the risk companies take when they tout the effectiveness of their leak detection technology.

Just 10 days before the accident, Enbridge Inc., which operates the Michigan pipeline, told federal regulators it could remotely detect and shut down a rupture in eight minutes. But when the line burst open, it took Enbridge 17 hours to confirm the spill.

Pipeline operators use a variety of methods to look for leaks, but the remote leak detection system—a combination of sensors, gauges, computer software and control room technicians called controllers—is the only one that offers real-time, continuous monitoring along the length of the line.

Operators often cite these systems as an example of their dedication to pipeline safety, particularly when they're questioned by citizens who fear that a leak may go undetected for hours or days.

Such questions are frequently asked in Nebraska, one of the six states along the Keystone XL's path. The line's southern leg, from Cushing, Okla. to the Texas Gulf Coast, is already under construction. The U.S. State Department is expected to decide early next year whether to approve the northern leg, which would cross the U.S.-Canada border.

Ninety miles of the Nebraska section of the line is scheduled to pass through the Ogallala/High Plains aquifer, which supplies drinking water to eight states and provides 30 percent of the groundwater used for irrigation nationwide. Twenty miles of that section will be buried in an area where the water table is less than 20 feet beneath the surface. An additional 70 miles will cross areas where the water table is 20 to 50 feet below ground.

Residents of Holt County, Neb., feel particularly vulnerable, because the region's high water table combined with the loose, sandy soil means any spilled oil would move quickly into the aquifer. Most residents get their drinking water from shallow private wells that aren't tested regularly for contaminants.

Landowners elsewhere along the route have similar concerns.

Dwayne and Zona Vig raise drug-free lean beef on a 15,000-acre ranch in Meade County, S.D., where the pipeline would be buried in the same field as an existing water line. The Vigs are especially worried about leak detection and emergency response. Their ranch is accessible only by dirt roads that are impassable during heavy rains, and they live 100 miles from the nearest hospital.

Zona Vig fears a small oil leak could go undetected for days, especially if it spread underground without reaching the surface.

"That is the one that scares us," she said.

Large Spills Easier to Detect

Kuprewicz and other experts say the reason remote systems find so few leaks is fairly simple: Remote sensors are good at detecting large spills and ruptures, but they're not so good at detecting smaller spills, which are far more common on the nation's pipeline system.

"Leak detection systems are imperfect," said Andrew Black, president of the Association of Oil Pipelines, which represents pipeline owners and operators. "...I think all operators will acknowledge that large ruptures are easier to detect."

In most cases, a well-designed, computer-based system will "find a major rupture in much less than 10 minutes," said Randy Allen, a staff consultant at UTSI International, which specializes in pipeline automation and leak detection. But Allen also pointed out that some smaller, slower leaks are virtually impossible to detect remotely.

According to InsideClimate News' analysis of PHMSA data, 76 percent of the leaks between 2002 and July 2012 involved fewer than 30 barrels of oil (1,260 gallons).

The agency's database contains the most extensive pipeline spill data available to the public and includes every accident larger than five gallons. It recorded a total of 1,763 oil pipeline spills in the 10-year period.

The entries for almost half of the spills—803—did not identify how the leak was detected, in part because PHMSA has less stringent reporting requirements for leaks between 5 gallons and 5 barrels (210 gallons) in size. So InsideClimate News confined its analysis to the remaining 960 spills.

Black said that if InsideClimate News narrowed its analysis to the larger incidents, it would find the percentage of leaks detected by sensors to be much higher. And he was right—to a point.

PHMSA considers all spills greater than 50 barrels (2,100 gallons) to be "significant." But to test Black's hypothesis, InsideClimate News studied the data for spills that would be considered highly significant—those larger than 1,000 barrels (42,000 gallons).

This time the data showed that remote sensing systems detected 20 percent of the spills, a big improvement over the 5 percent detected in our original analysis.

Yet the general public discovered almost as many spills—17 percent—as the sensors. And 42 percent were discovered by employees at the scenes of accidents.

"The fact that 80 percent of leaks larger than 42,000 gallons go undetected by [remote] leak systems is a real sign of a problem," said Swift, the NRDC attorney.

Monitoring the Keystone XL

TransCanada has told federal and state regulators that the Keystone XL's leak detection system will be able to detect spills below 1.5 percent of the pipeline's flow.

Allen said one to two percent is "the most anyone's going to guarantee, because there are some hydraulic behaviors that thwart perfect leak detection."

He also said TransCanada's system is considered to be among the best in the industry, and he believes there is "a reasonable chance" that the company may be able to beat the 1.5 percent limit on some segments of the Keystone XL. "But they're not going to tell you they can, because they're not sure."

Because the Keystone XL will carry so much oil, that 1.5 percent represents hundreds of thousands of gallons per day.

The 36-inch wide pipe will be one of the largest pipelines in the country, with an initial capacity of 700,000 barrels per day that can later be expanded to 830,000 barrels—nearly 35 million gallons—per day.

A spill involving 1.5 percent of the initial capacity would be 10,500 barrels, or 441,000 gallons a day.

When calculated for the expanded capacity, that 1.5 percent comes out to 12,450 barrels, or 522,900 gallons a day.

Allen says TransCanada can use a technique called static pressure testing to look for smaller leaks.

But to do that, an operator must be willing to periodically shut down the line—and interrupt its business—to conduct the tests.

TransCanada declined to make a technical expert available for interviews. Semmens, the TransCanada spokesman, said the company will run static pressure tests whenever the line is shut down due to "operational constraints," such as a temporary delay in scheduled deliveries.

The Risk of False Alarms

Most pipeline companies buy their leak detection systems from specialized engineering firms, then customize the systems to meet the geographic and technical needs of individual projects.

Yet they all basically operate the same way. Sensors along the pipelines measure temperature, pressure, flow rates and other hydraulic data. The information feeds into the control room, where it serves two functions—tracking the amount of oil delivered to refineries and other customers, and monitoring the pipeline for potential leaks.

When the leak detection software senses something that could be a leak—perhaps an abrupt change in pressure and flow rates—it triggers an alarm. The controllers then analyze the data to determine whether there's really a problem.

This last step is crucial, because many alerts turn out to be false alarms. For instance, column separation—what's essentially a large bubble in the flow of oil—can look just like a leak on the remote systems.

False alarms can lead to costly mistakes.

At the time of the Michigan spill, Enbridge's controllers were working 12-hour shifts and simultaneously monitoring data coming in from multiple pipelines. When pipeline 6B ruptured, 16 alarms went off. But the controllers and analysts concluded they were false alarms caused by column separation, and it was 17 hours before they realized they had a spill.

Operators can increase the sensitivity of their remote sensing systems to identify smaller leaks, Kuprewicz said. But when they do that, they also increase the chances of setting off false alarms.

"If you get a thousand [false alarms] a month, what happens when you get a big [real] one?" Kuprewicz said. "How do you tell the difference? You can't."

Allen, the UTSI consultant, said experienced controllers can recognize the warning signs even before the system sounds an alarm. But because they're usually busy making sure batches of oil are delivered to the right destinations, they're not necessarily looking at the hydraulic data used for leak detection.

No System is Perfect

The industry spends millions of dollars a year trying to improve remote leak detection, but Kuprewicz said there are limits to what can be done.

Leak detection systems work best for simple pipelines where the oil is flowing at a steady rate. But if a pipeline is shut down, or if the flow rate keeps changing—as is common with most pipelines, including the Keystone XL—detecting a leak is more difficult, because it's hard to determine how much oil should be in the pipeline at any given time.

Operators also rely on ground and aerial patrols to detect the smaller spills. Semmens, the TransCanada spokesman, said the company will conduct ground or aerial surveys on the Keystone XL at least once every 2 weeks. An Enbridge spokesman said his company will follow a similar schedule on the new pipeline it is building in Michigan, to replace the one that ruptured in 2010.

But even visual surveys aren't foolproof, because some spills never reach the ground surface, Kuprewicz said. That's why the people who patrol pipeline right-of-ways are always on the lookout for dead vegetation, a possible sign of an underground leak.

Another option is to install external sensors that can detect leaks smaller than 10 gallons per day. But these sensors are expensive and are rarely used.

The bottom line, said Kuprewicz, is that there's no perfect solution to spotting oil spills. Ideally, companies should combine the best leak detection technology with experienced operators—but even then, some leaks will go undetected.

"No one sells leak detection systems claiming they will not work," he said. "So you want to be careful about the claims and how realistic they are."

KEYSTONE XL WOULD NOT USE MOST ADVANCED SPILL PROTECTION TECHNOLOGY

It would cost less than $10 million—roughly 0.2 percent of the Keystone XL's budget—to add safeguards to protect the crucial Ogallala aquifer from spills.

By Lisa Song

In 1998, activists in Austin, Texas filed a lawsuit to protect their local aquifer from a proposed gasoline pipeline. By the time the project was built, the operator had been forced to add $60 million in safety features, including sensor cables that could detect leaks as small as three gallons a day. Some say the Longhorn pipeline is the safest pipeline in Texas, or perhaps the nation.

Now a much larger pipeline—the Keystone XL—is being proposed across the Ogallala/High Plains aquifer, one of the nation's most important sources of drinking and irrigation water. Yet none of the major features that protect Austin's much smaller aquifer are included in the plan. In fact, they haven't even been discussed.

The leak detection technology that will be used on the Keystone XL, for instance, is standard for the nation's crude oil pipelines and rarely detects leaks smaller than 1 percent of the pipeline's flow. The Keystone will have a capacity of 29 million gallons per day—so a spill would have to reach 294,000 gallons per day to trigger its leak detection technology.

The Keystone XL also won't get two other safeguards found on the 19-mile stretch of the pipeline over Austin's aquifer: a concrete cap that protects the Longhorn from construction-related punctures, and daily aerial or foot patrols to check for tiny spills that might seep to the surface.

Experts interviewed by InsideClimate News estimate it would cost less than $10 million—roughly 0.2 percent of the Keystone's $5.3 billion budget—to add external sensor cables, a concrete cap and extra patrols to the 20 miles of the pipeline in Nebraska where a spill would be most disastrous. The water table in that area lies less than 20 feet below the surface and provides ranchers with a steady supply of fresh water.

TransCanada, the company that wants to build the Keystone XL, says the project meets or exceeds federal pipeline standards. In June, Russ Girling, TransCanada's president and CEO, said it will be "the safest, most advanced pipeline ever built in North America."

Spokesman Shawn Howard said trained experts will monitor the pipeline 24/7 from a state-of-the-art control center. His colleague Grady Semmens said operators would shut down the pipeline within 10 minutes of detecting a problem.

TransCanada also has pledged to follow 57 conditions that it says exceed federal standards. That list doesn't include any of the three major safeguards that protect the Austin aquifer. And an analysis last year by the Natural Resources Defense Council found that most of the 57 conditions are identical to existing federal regulations.

"TransCanada applies industry best practices, many of which exist due to the potential lack of federal regulations, advances in technology, construction practices and methodologies from both a safety and quality perspective," Howard said in an email.

Environmental groups and landowners have been fighting the Keystone XL project for years, but the possibility of adding safeguards like those used on the Longhorn pipeline hasn't been part of the debate.

Most environmental groups want the pipeline stopped altogether, primarily because the Canadian crude oil it will carry has a much larger carbon footprint than conventional oil. NASA climate scientist James Hansen has famously called the pipeline a "fuse to the biggest carbon bomb on the planet."

Carl Weimer, executive director of the Pipeline Safety Trust, said many of the protective measures used on Longhorn could also improve safety on the Keystone XL, as well as other pipelines that cross vulnerable lands. His nonprofit, nonpartisan group has spent years advocating for stronger federal pipeline construction and safety rules. "We just assume pipelines will end up in a lot of places, so let's just make them as safe as possible."

The industry expects to build or repurpose more than 10,000 miles of pipelines over the next five years to transport heavy crude from Canada's oil sands region.

Weimer said it often takes years to change even a minor regulation, because the rule-making process is slowed by the "big gorilla in the room"—industry representatives who are part of the process and are reluctant to adopt changes that could impact their bottom line.

"Cost is the major factor," said Mohammad Najafi, a civil engineering professor at the University of Texas-Arlington and editor-in-chief of the *Journal of Pipeline Systems Engineering and Practice*. Najafi declined to comment on the Keystone XL, but said that, in general, operators "that don't take extra measures do so because they're private companies with investors, and they cut costs as much as they can."

But Najafi warned that increased spending doesn't automatically boost pipeline safety, because resources could be wasted on badly designed technologies.

Najmedin Meshkati, a University of Southern California civil engineering professor who studies workplace safety culture, said even the best technology can't guarantee safety. "The human factor is really where the rubber meets the road....No piece of hardware can replace [a] good safety culture."

How the Austin Aquifer Got Extra Protections

Longhorn opponents had two major advantages in their lawsuit to protect Austin's aquifer: money and popular opinion.

The lawsuit was filed by the city of Austin, the Barton Springs/Edwards Aquifer Conservation District and several landowners. But most if not all of the funding came from Holly Corp., a Texas refinery owner that saw Longhorn's gasoline pipeline as a competitive threat to its business. (Longhorn later sued Holly for antitrust violations. Holly then countersued alleging unfair competition.)

Without Holly's money, it would have been difficult for the pipeline opponents to finance their battle against Longhorn—a partnership of oil giants BP Amoco, Exxon and other companies. Renea Hicks, who represented the Conservation District in the case, said Holly's funding allowed the opponents to hire top-notch experts and to persist for the three years it took to settle the suit.

The case also had wide support among the politically active and environmentally conscious residents of Austin, a city of 600,000 at the time and a liberal enclave in a largely conservative state. The fuels carried by the Longhorn are highly flammable and spread quickly when spilled into water. Not only does the Edwards aquifer supply drinking water to the Austin area, it also feeds a popular swimming hole in the city's Zilker Park.

"Everyone got involved," Hicks said. "The purity of that [spring] is kind of a symbol in Austin."

Kirk Holland, a geologist and general manager of the Conservation District, said that in the end, Longhorn "essentially had to make those extraordinary commitments in order to operate [in this area]. That pipeline is the best-protected, most monitored pipeline in Texas—and maybe the nation."

"Keystone XL deserves more, probably, in my personal opinion," Holland added.

Although the Keystone XL would run 1,200 miles across the nation's heartland, the area of greatest concern has always been Nebraska, where it crosses 222 miles of the Ogallala aquifer. Last year, TransCanada agreed to move the line out of Nebraska's Sandhills region, a fragile landscape that became a symbol for the project's opponents. But the new route still goes through 20 miles of Nebraska where the water table is less than 20 feet below ground—high enough for groundwater to bubble to the surface during the spring. An additional 70 miles crosses areas where the water table is 20 to 50 feet underground.

Last year, just weeks before TransCanada agreed to the Sandhills reroute, the company offered to build a concrete containment structure around a pump station planned for a sensitive area of Holt County. It also agreed to post a $100 million bond to be used if the company failed to clean up an oil spill in the Sandhills. TransCanada withdrew both offers after the pipeline was rerouted.

"Those commitments were specific to the area that went through the Nebraska Sandhills where the aquifer was at or near the surface," said Howard, the company spokesman. "Since the new route will go through an area that is not part of the defined Nebraska Sandhills, those measures are not required."

The federal Pipeline and Hazardous Materials Safety Administration (PHMSA) requires operators to follow more stringent rules in High Consequence Areas, or HCAs, which are considered especially vulnerable to the effects of an oil spill. Pipelines that could affect HCAs are built with thicker walls and the insides are inspected at least once every five years.

Najafi, the University of Texas engineer, said the entire aquifer under Nebraska should qualify as an HCA.

"Nebraska is a sensitive area, and they need to treat it like that," he said. "In Nebraska, they need extra measures to protect the water. We can't live without water."

PHMSA doesn't release the locations of HCAs to the public due to security reasons, so it's unclear how much of the Ogallala aquifer falls within an HCA.

According to Howard, only three miles of the route in Nebraska crosses an HCA. But he said that TransCanada would exceed PHMSA requirements by running an inspection device through the entire pipeline, including areas outside HCAs.

Little Known About How a Spill Might Affect the Aquifer

The Keystone XL was originally supposed to run from Alberta, Canada to the Texas Gulf Coast. But in January 2012 the Obama administration turned down TransCanada's application for the State Department permit it needed to cross the U.S.-Canada border, and TransCanada split the project in two. The segment from Cushing, Okla. to Texas, which did not need a federal permit, is already under construction. A decision on the northern segment is expected in early 2013.

Much of the opposition to the Keystone XL has focused on the type of oil it would carry: Bitumen is a particularly heavy form of crude oil extracted from Canada's tar sands region. It is so thick that it can't flow through pipelines until it's diluted with liquid chemicals to form what's known as diluted bitumen or **"dilbit."**

The same federal standards that apply to gasoline and crude oil pipelines also apply to pipelines carrying dilbit—even though dilbit doesn't behave like conventional crude oil when it spills into water.

The nation's first major dilbit spill occurred in July 2010, when a ruptured pipeline released a million gallons of dilbit into the Kalamazoo River. As the light, liquid chemicals in dilbit began evaporating, the heavy bitumen sank into the river. Nearly two and a half years later, Enbridge Inc., the Canadian company that owns the ruptured pipeline, is still struggling to clean up the Kalamazoo. The current cost of clean up exceeds $800 million.

The Enbridge accident showed what happens when dilbit spills into a river. But little is known about how dilbit might behave in an aquifer, said Wayne Woldt, a University of Nebraska-Lincoln professor who studies groundwater management.

Once an aquifer is contaminated, it's virtually impossible to restore it to its pristine condition, Woldt said. The extent of damage would depend on the size of the spill and on how the dilbit moves within the aquifer.

"Some say it would pollute the Ogallala aquifer a tremendous amount. Others say it wouldn't be a big deal. I don't know, because I haven't found the research that would answer this question," Woldt said. "I think we're all operating in a vacuum of information."

Woldt has tried without success since June 2011 to secure funding for a study modeling the effects of dilbit on the Ogallala aquifer.

Most Sensors Detect Only Major Pipeline Ruptures

TransCanada ran newspaper ads in Nebraska last summer assuring residents that the Keystone XL will be protected by 21,000 sensors that relay information to the company's control center once every five seconds. But a recent examination of PHMSA data by InsideClimate News showed that between 2002 and July 2012, only five percent of U.S. crude oil pipeline spills were detected by leak detection systems like those planned for the Keystone XL. Those systems detect major pipeline ruptures, not the "weeps and seeps" that can accumulate into large spills. That means the Keystone system is unlikely to detect spills smaller than hundreds of thousands of gallons per day.

The technology used on the Longhorn pipeline is more sophisticated. In addition to a standard leak detection system, hydrocarbon-sensing cables are attached to the pipeline's exterior, where they can trigger an alert if the pipeline's contents drip out. According to Longhorn, they can detect spills as small as three gallons a day.

Because external sensors are more expensive, they are used on less than one percent of the nation's oil pipelines. They're usually found on small stretches over sensitive river crossings, aquifers and other areas where an oil spill could be disastrous.

PHMSA requires operators to "have a means to detect leaks" on their pipelines, but it sets no standards for how effective the systems must be. The agency is in the midst of a two-year study on leak detection, but it could be years before the results are incorporated into regulations.

Weimer, the Pipeline Safety Trust director, said that while PHMSA supports advances in pipeline technology, the agency "is hesitant to ever tell the industry what kind of system they need to use." As a result, it takes a long time for new safety technologies to percolate through the industry.

Longhorn purchased its sensor cables from Houston-based Tyco Thermal. Ken McCoy, general manager of the group that designed the cable, said his product costs more per mile than the typical leak detection technology used for oil pipelines. He refused to compare prices directly but estimated that installing his company's sensors on 20 miles of the Keystone XL would raise the project's $5.3 billion price tag by less than 0.13 percent. He said that price includes the added cost of adapting the technology for Nebraska's high water table.

TransCanada has no plans to add external sensors. Howard, the company spokesman, said external sensing technology "is not the best method to use" on the Keystone XL, because it is "subject to localized conditions such as water table and soil conditions and therefore reliability and maintenance can be an issue."

TransCanada's concerns are valid but not insurmountable, said Richard Kuprewicz, a pipeline safety expert who is president of the consulting firm Accufacts, Inc. Kuprewicz has worked with TransCanada in the past, but is not involved with the Keystone XL.

TransCanada could choose from a variety of external sensors, including fiber optics or acoustic sensing technology, Kuprewicz said. Investing in these technologies might make good business sense if concerns about oil spills are holding up a multi-billion dollar project, he said. "If you can't solve a problem, either reroute the line, or come up with a solution that has a high degree of expectation it'll do its job."

Howard said TransCanada is involved in an industry research group that has undertaken a "multi-year effort to identify and quantify improved capabilities for the detection of small leaks," and that the design of Keystone XL "positions us well to leverage evolving technologies in the future."

Other safeguards on the Longhorn—including concrete caps and more frequent foot and aerial patrols—do not involve expensive technology.

Most of the 700-mile Longhorn pipeline route is inspected once a week, but the 19-mile section over Austin's aquifer is inspected every day.

TransCanada will meet the federal requirement of at least 26 inspections per year for the Keystone XL. According to Howard, aerial surveys will be conducted about once every two weeks, along with "'on foot' inspections [that] occur more frequently as part of our operators' regular tasks and routine."

The Keystone XL will not include a concrete cap like the one used for Longhorn.

Weimer said concrete caps could prevent excavation damage, a leading cause of serious pipeline accidents. But the cap may be of limited use along the Keystone, he said, because most of the pipeline runs through sparsely populated regions that are unlikely to see much construction activity.

The Longhorn pipeline is now owned by Magellan Midstream Partners, which plans to convert the line to carry crude oil. A Magellan spokesman said the company will continue to use and maintain all of Longhorn's safety features.

Keystone Protest Continues

On Feb 17, Keystone XL opponents will gather for their fourth large protest outside the White House in hopes of persuading President Obama to stop the project. For now there is little talk about what they will do if the pipeline is approved. Anthony Swift, a policy analyst for the Natural Resources Defense Council, said it's "premature" to discuss whether his organization would push for stronger pipeline protections if that happens.

As the debate over the project continues, the National Academy of Sciences is analyzing existing research studies to determine whether dilbit corrodes pipelines more quickly than conventional crude oil. But the National Academy report isn't due until next summer, and by then it will be too late for it to have much impact on the construction of the Keystone XL. The administration is expected to make its decision in early 2013, and TransCanada has said it is prepared to begin work immediately.

LITTLE OVERSIGHT FOR ENBRIDGE PIPELINE ROUTE THAT SKIRTS LAKE MICHIGAN

Despite calls for extra protection for the vital watershed, Enbridge expects to have its final permits from Indiana to begin construction in May or June.

By David Hasemyer and Lisa Song

In the northwestern corner of Indiana a major pipeline project is planned that will carry vast quantities of heavy Canadian crude oil across four rivers that flow into Lake Michigan, where 10 million people get their drinking water. The pipeline will cross one river just 11 miles from the lake. It crosses the other three rivers less than 20 miles from the lake.

Because the pipeline runs so close to Lake Michigan—and because it is being built by a company with a history of pipeline spills in the region—a growing coalition of environmental groups is demanding that it be given extraordinary oversight and protection.

But getting those protections will be almost impossible.

No federal or Indiana agency has authority to require the pipeline's Canadian operator, Enbridge, Inc., to move the line out of the Lake Michigan watershed—or to add extra safeguards, including sophisticated technology that can detect even minor spills.

Enbridge was responsible for the most expensive oil pipeline spill in U.S. history, a 2010 rupture near Marshall, Mich. that dumped more than a million gallons of Canadian crude into the Kalamazoo

River, a Lake Michigan tributary. Oil from Line 6B contaminated about 36 miles of the river before cleanup workers managed to stop it roughly 70 miles from the lake. Enbridge was fined $3.7 million for breaking more than 20 federal rules, and the National Transportation Safety Board reprimanded the company for "a complete breakdown of safety."

Enbridge's new project will replace Line 6B with a larger pipe that can carry as much as 21 million gallons per day, more than double its capacity when it fouled the Kalamazoo. Like the existing 6B, it will carry a thick oil from Canada's tar sands called bitumen that has been thinned with liquid chemicals to form diluted bitumen, or dilbit. During the 2010 spill, the light chemicals evaporated while the bitumen slowly sank, leaving a mess that is still being cleaned up today. The U.S. Environmental Protection Agency told Enbridge in October that submerged oil needs to be dredged from 100 acres of the river.

"There have got to be lessons learned from the Kalamazoo spill," said Steve Hamilton, a Michigan State University professor of ecology and environment who served on the EPA team that recommended the dredging. "There are legitimate issues to be concerned about in Indiana."

A major spill into one of the Indiana rivers would be even more disastrous than the Michigan spill, the environmental groups say, because the pipeline's crossing points are so much closer to Lake Michigan.

"If the Marshall spill would have happened here, it would have been in Lake Michigan," said Nathan Pavlovic, a land and advocacy specialist with Save the Dunes, a 60-year-old nonprofit dedicated to protecting the Indiana Dunes and the Lake Michigan watershed. "There are just too many unanswered questions at this point to consider this a safe project when you consider the devastating consequences of a spill."

In addition to serving as one of the nation's most important drinking water supplies, Lake Michigan supports recreational activities that are vital to the regional economy and an ecosystem that is home to rare plants and animals.

Erin Argyilan, a geoscientist who has lived near the lake for most of her life, wants Enbridge to disclose how it would respond to spills in various parts of the lake's watershed. An oil spill in a wetland, for instance, would behave very differently from oil spilled into a fast-moving tributary.

"Where would it be after two hours? After four hours? After six hours?" asked Argyilan, who sits on the Save the Dunes board and chairs the geosciences department at Indiana University Northwest. "This type of modeling [would] put a realistic face on what could happen.

"It's unfortunate to always go to the worst case scenario, but in this case it's necessary when one of the world's largest fresh water resources is at stake."

Argyilan questioned Enbridge about its plans at a public meeting in September, but she said she got only general assurances that response teams were in place and that shut-off valves would quickly close the pipeline if a spill occurred.

Pipeline operators aren't required to file their emergency response plans with federal regulators until their projects are built. And they're not required to share them with the public. Local emergency officials in Indiana told InsideClimate News they trust that Enbridge's emergency plans for the existing 6B are sufficient—even though some said they hadn't reviewed those plans.

Save the Dunes isn't trying to stop the 6B replacement project. The 43-year-old line is an essential part of Enbridge's Lakehead System, which transports as much as 75 percent of the crude oil consumed by refineries in the Upper Midwest. Without the pipeline, many would have trouble getting the large quantities of crude they need to produce jet fuel, gasoline, heating oil and other products critical to the region's economy.

What the group is demanding, however, is that state and federal agencies use whatever authority they have over the project to require additional safeguards.

In a statement last week signed by the National Wildlife Federation and several other groups, Save the Dunes urged the Indiana Department of Environmental Management "to ensure that Enbridge implements every possible precaution to protect the people and natural resources of Northwest Indiana and Lake Michigan."

"We are questioning the wisdom of permitting a pipe that will dramatically increase the quantity of tar sands flowing through our region without quite a bit more due diligence," Pavlovic said.

In email responses to questions from InsideClimate News,

Enbridge said the 6B project meets all the standards set by the federal Pipeline and Hazardous Materials Safety Administration (PHMSA), which regulates the nation's pipelines. Spokesman Larry Springer said the company will exceed some of those standards, including conducting internal inspections along the entire 285-mile project, which runs from Griffith, Ind. to Sarnia, Ontario.

"All valves installed on Line 6B will have remote control capability, which is not a PHMSA requirement," Springer said. "In addition, we X-ray or ultrasonic test 100 percent of the welds during pipeline construction, which exceeds the current PHMSA requirement of 10 percent."

Lack of Additional Safeguards

Enbridge's plans for the new 6B do not include installing some safeguards that are readily available but are not required by federal law.

InsideClimate News reported earlier this month that pipeline operators can improve leak prevention and detection by capping their lines with concrete or by adding more aerial or foot patrols.

They can also install external sensors that are far more sensitive than the leak detection systems found on most of the nation's pipelines. Those systems rarely find leaks until several thousand gallons of oil a day have been spilled. The more expensive external sensors can detect leaks as small as three gallons a day.

Enbridge doesn't use external sensors on any of its pipelines. Spokesman Graham White said the company is studying the technology and may use it on future projects. But the study won't be done in time for the 6B replacement.

Enbridge is Canada's largest transporter of crude oil, with a 2011 operating income of more than $1 billion. The company plans to spend $8.8 billion on pipeline projects in the United States over the next several years, including $1.3 billion on the 6B replacement project.

Line 6B begins in Griffith, Ind., where Canadian oil is held in a large Enbridge facility. It then travels east across 60 miles of Northern Indiana, crosses into Michigan and finally into eastern Canada.

Work has already begun on the Michigan leg of the project,

despite protests and court challenges from landowners.

Springer, the company spokesman, said Enbridge expects to have its final state permits from Indiana in time to begin construction in May or June 2013.

Indiana Doesn't Control Pipeline Route

Control of 6B's route through Indiana rests almost entirely with Enbridge.

The U.S. Environmental Protection Agency, the Fish and Wildlife Service and the Army Corps of Engineers ensure that wildlife, waterways and wetlands are protected during construction. And PHMSA makes sure pipelines are built and operated according to government safety standards.

But regulating the location of pipelines is left to the states, and no state agency in Indiana has been charged with that responsibility. This means Enbridge can decide where to run its pipeline with little oversight.

Carl Weimer is executive director of the Pipeline Safety Trust, a nonprofit, nonpartisan watchdog organization based in Bellingham, Wash. He said Indiana's situation isn't unusual, because many states don't control pipeline routes within their borders.

"There is a disconnect between the siting of a pipeline and pipeline safety issues," Weimer said. "It's common for state and local agencies to defer to PHMSA. So when PHMSA kind of nods its head that is taken as validation the siting matter has been addressed."

The state of Nebraska found itself in a similar predicament when TransCanada, Canada's largest pipeline operator, proposed building the Keystone XL pipeline through the Ogallala aquifer, which provides drinking water for eight states and 83 percent of Nebraska's irrigation water.

Despite a public outcry over the route, state regulators had no authority to change it. Finally, TransCanada moved the route out of the most fragile area, known as the Sandhills, but residents are still fighting the project, which is still waiting for approval from the U.S. State Department. In 2011, Nebraska passed a law allowing the state's Public Service Commission to evaluate pipeline routes, but the new law applies only to pipelines built after the Keystone XL.

In Indiana, the Utility Regulatory Commission regulates intrastate natural gas pipelines, but not oil lines. The Indiana Division of Natural Resources focuses on minimizing the impact of construction projects on water, plants and animals and the navigability of the state's rivers.

Two of the last state permits Enbridge needs to start work on 6B are from the Indiana Department of Environmental Management (IDEM), which has authority to make limited routing changes when construction affects wetlands and water crossings. Agency spokesman Robert Elstro said IDEM can place conditions on the permits to ensure compliance with state water pollution laws and regulations.

Save the Dunes is pressuring IDEM to use that authority. In the statement it released last week, the group made two specific requests: That IDEM require Enbridge to pay for independent monitors during the line's construction, and that it force the company to reroute the line around sensitive wetlands.

Enbridge is already familiar with independent monitors. The state of Wisconsin required the company to hire them when Enbridge built an oil pipeline across Wisconsin several years ago. Between 2007 and 2008, the monitors documented more than 500 violations of state environmental rules, and Wisconsin fined Enbridge $1.1 million.

"It speaks to the critical nature of these independent monitors," Pavlovic said. "Without [them], those violations probably would have gone undocumented and undetected."

Rerouting the pipeline around sensitive wetlands also makes sense, Pavlovic said, particularly near Hudson Lake in LaPorte County.

The line is currently routed north of the lake. But Pavlovic said it should be moved south of the lake, where there are far fewer wetlands and where a natural gas pipeline owned by an Enbridge subsidiary is already in the ground. The gas pipeline was built there about a decade ago, after a federally mandated environmental study concluded that the land south of the lake was less vulnerable than the land in the north.

The change would add about a half mile to the line's 285-mile route.

"This is a variation that has been assessed in the past, and [the southern route] was clearly shown to be preferable," Pavlovic said.

Springer, the Enbridge spokesman, told InsideClimate News that the company will "follow any route alternatives" IDEM proposes and "would consider hiring independent monitors."

Emergency Managers Trust Enbridge Response Plans

Since its 2010 Kalamazoo pipeline spill, Enbridge has taken steps to make its pipelines safer, most of them in response to a Corrective Action Order issued by PHMSA.

The company has developed better tools and technology for worst case waterborne spills, increased spending on pipeline integrity management, added new emergency training programs for employees and will spend $50 million between 2012 and 2013 to improve "equipment, training and overall response capabilities," Springer said. This year Enbridge held a company-wide emergency response drill in Houston, where it simulated an oil spill into a large body of water.

Emergency managers in three of the Indiana counties the pipeline will cross told InsideClimate News they are satisfied with the company's emergency response plans for the current 6B and believe the new pipeline will be safe. Although they haven't been provided response plans tailored specifically for the unique conditions a dilbit spill would create, they said they have universal procedures that can cover a multitude of scenarios.

"You have to ask yourself the question: 'Is it [a spill] going to happen?'" said Russell Shirley, director of the Department of Emergency Management in Porter County. "Anything is possible, but it is probable?

"I don't think it's probable."

The emergency managers said they get most of their information from an annual meeting held by about two-dozen pipeline companies, including Enbridge. The meeting satisfies a federal requirement that pipeline operators make representatives available to local officials.

Shirley said he hasn't seen Enbridge's response plan, hasn't met with an Enbridge representative and has never engaged in any drills with Enbridge. But he said his 20-member hazmat team could quickly confront a spill and that he could call 40 additional hazmat responders from adjoining counties.

Jeff Hamilton, director of LaPorte County's hazmat department, said that outside of the annual meeting, he has only sporadic contact with Enbridge, although a company liaison is always available for calls.

Hamilton said Enbridge has assured him that it can quickly marshal a force of contractors to go to work on a spill.

"We know what kind of support we can get from them and how quickly we can get it," he said.

On one point he is supremely confident: "We can keep it held out of Lake Michigan."

In Lake County, Elijah Cole Jr., deputy director of the Emergency Management Agency, said he has never seen Enbridge's response plans. He thinks they are filed away in the director's office somewhere.

"I'm sure we have them," he said. "The pipeline companies are supposed to give them to us every year."

Cole said his agency depends on Enbridge and other pipeline operators to react to any spills their companies may cause.

"Each one of the pipelines has their own plan," he said. "They know what's in them. The pressure. They know the circumstances of what's going on. So they would have all of that information. It would be their responsibility."

NEW PIPELINE SAFETY REGULATIONS WON'T APPLY TO KEYSTONE XL

Proposed federal rules to strengthen pipeline safety won't be in place before construction could begin on the Keystone XL or other new dilbit pipelines.

By Elizabeth McGowan and Lisa Song

WASHINGTON—Efforts to beef up oversight of the nation's oil pipelines are progressing so slowly that it's unlikely any additional safeguards will be in place before construction begins on thousands of miles of new pipelines, including the proposed Keystone XL.

Part of the delay stems from how slowly the Pipeline and Hazardous Materials Safety Administration (PHMSA)—the federal agency with the authority to issue new regulations—is moving on its rulemaking process. For instance, PHMSA began examining at least six safety regulations in October 2010, three months after a ruptured pipeline spilled more than 1 million gallons of oil into Michigan's Kalamazoo River. None of those changes is in effect nearly two years later.

Congress's latest pipeline safety bill, which was signed into law in January, did little to speed up the process.

The measure did not address two of the key regulatory failures that InsideClimate News found during a recent seven-month investigation of the Michigan spill. It did not force PHMSA to enforce deadlines for repairing pipeline defects or require that pipeline operators identify exactly what type of oil is flowing through their lines. Both of those failures were also detailed in a report released this month by the National Transportation Safety Board.

"Tens of thousands of miles of new pipelines are going into the ground, and there aren't going to be regulations that make them safer for years," said Carl Weimer, executive director of the Pipeline Safety Trust, a nonprofit watchdog organization based in Bellingham, Wash.

Representatives of the Trust testified at least 10 times on Capitol Hill as Congress was shaping the Pipeline Safety Act.

"We saw that the final bill really didn't do much for safety," Weimer said. "We're just happy it didn't go in the wrong direction. With this Congress, not going in the wrong direction is a win."

The bill did address two problems that became apparent after the Michigan disaster. It authorized a study of diluted bitumen, or dilbit—the type of oil that spilled into the Kalamazoo and would also be carried on the Keystone XL. And it ordered the Department of Transportation to study the technology that the pipeline industry uses to detect leaks. PHMSA is a division of the Transportation Department.

Neither of those studies will be done in time to have much impact on the new pipeline construction that is predicted for the United States.

The dilbit study won't be ready until next summer, and it will consist only of a review of the existing literature, not new research. The leak detection study won't be ready until 2014 at the earliest, because Congress stipulated that PHMSA spend two years on the project.

Democrat John Dingell, who has represented southeastern Michigan in the House for 58 years and helped craft the House version of the bill, called it a "good first step" but said "we have much more to do."

"The NTSB report on the Enbridge spill in the Kalamazoo River highlights important issues which Congress and PHMSA need to address to ensure that our aging pipeline system is as safe as possible," the former chairman of the Energy and Commerce Committee said via e-mail.

Rep. Fred Upton (R-Mich.) the current committee chairman who crafted the House version of the bill with Dingell, didn't return requests for comment. Neither did Sen. Frank Lautenberg (D-N.J.) who helped to shepherd the bill through the Senate.

Former Rep. Jim Oberstar (D-Minn.), who was voted out of office in 2010 several months after he chaired a House Transportation and Infrastructure Committee hearing on the Marshall spill, said the bill was so weak that he wouldn't have supported it. It's up to regulators to intervene when pipeline operators ignore the public's health and safety, he said, and Congress fell short this time.

"If you don't have a corporate culture of safety, then the public sector must rigorously oversee operations where there is a hazard to public health and safety," Oberstar said. "And that is the issue we're dealing with in the pipeline sector."

Sara Gosman, a lecturer at the University of Michigan Law School who studies pipeline safety in the Great Lakes region, said that instead of nibbling around the edges, Congress should tackle pipeline safety in the same overarching manner that the landmark Clean Air and Clean Water acts focus on protecting people and natural resources.

"We haven't had a big environmental act passed in this country in the last 20 years," Gosman said. "We're just not seeing shifts in the way pipelines are regulated. The opportunity is there ... but Congress just isn't looking forward."

Below is a review of six regulatory problems that became apparent after the Michigan pipeline accident and the action that is—or isn't—being taken in response.

Pipeline Contents Still a Mystery

The federal officials and cleanup crews who rushed to the scene of the Marshall, Mich., accident didn't know for at least two weeks that they were dealing with dilbit, not conventional oil. Current regulations don't require operators to provide that information, even after a spill.

The Kalamazoo disaster showed that the distinction is important. Conventional oil floats on the water's surface, where it can easily be vacuumed or skimmed away. But bitumen is so thick that it must be thinned with liquid chemicals before it can flow through pipelines. When the pipeline ruptured in Michigan, those light chemicals began to evaporate, compounding the concerns of health officials. The heavy bitumen then sank to the river bottom, making traditional cleanup methods almost useless.

The bill Congress passed didn't ask PHMSA to require pipeline operators to reveal whether their lines are carrying dilbit or conventional oil. But the NTSB has twice recommended that PHMSA direct operators to inform local emergency responders about the contents of their pipelines— last year after it investigated the 2010 San Bruno, Calif., gas line explosion that killed eight people, and earlier this month after it announced its findings for the Michigan spill.

Although Congress didn't mandate disclosure of pipeline contents, PHMSA has the authority to take that step itself. But PHMSA spokesman Damon Hill said no efforts are underway to do that.

Little is Known About Dilbit

Cleanup and health experts struggled to respond to the Michigan spill in part because so little is known about dilbit. InsideClimate News found few peer-reviewed articles on dilbit while researching the Michigan spill. Studies may have been conducted by the oil industry, but they're not available to the public. InsideClimate News relied on information from government publications, petroleum engineering textbooks and interviews with oil analysts, watchdog organizations and university scientists who have worked with the industry.

Although the Pipeline Safety Act directed PHMSA to conduct a study of dilbit, the study will be limited to a survey of the current scientific literature. No new research is planned.

A spokeswoman for the National Academy of Sciences, which is conducting the study for PHMSA, said it's too early to know whether it will include only peer-reviewed research or whether it will also look at industry and government publications.

The study mandated by Congress has also been limited to a narrow topic: whether dilbit is more likely than conventional oil to corrode pipelines. It will not explore two questions that emerged after the Michigan spill: How does dilbit differ from conventional oil when it spills into water? And how does that difference affect health and cleanup responses?

"Diluted bitumen behaves differently, particularly in water bodies [after] a spill," said Anthony Swift, an attorney and pipeline specialist with the Natural Resources Defense Council. "Spill responders haven't developed methods to contain and remediate those spills, and emergency response plans certainly don't incorporate the unique properties of dilbit in their response ... and nobody's evaluating that right now."

"We need regulatory agencies to do the [studies], or to commission it to be done. And it hasn't. In the end ... I would expect [the committee] to come to the same conclusion we [at NRDC] have, which is that there hasn't been enough basic science to understand the risks of moving diluted bitumen in pipelines."

The 12-member research committee from the National Academy of Sciences met for the first time Monday in downtown Washington. Swift, one of several experts invited to the meeting, was the only invited speaker from an environmental organization. He presented evidence along with representatives from Enbridge, TransCanada, the American Petroleum Institute, the Association of Oil Pipelines and researchers from the Alberta government.

The National Academy's study isn't expected to be finished until summer 2013. By then, the next president will almost certainly have decided whether the northern half of the Keystone XL pipeline can be built.

Deadlines for Repairing Corrosion and Other Defects Still Loose

The defect that led to the Michigan spill in 2010 was identified as early as 2005, when Enbridge, the pipeline operator, self-reported the anomaly to federal regulators. However, the company was allowed to delay the repairs without violating any PHMSA regulations.

Congress did not address the subject of repair deadlines in its legislation and it isn't on the agenda for PHMSA's current rulemaking session.

Any rules PHMSA proposes are reviewed by the agency's 15-member Liquid Policy Advisory Committee, made up of five representatives from government, five from industry and five from the public.

PHMSA is supposed to conduct a cost-benefit analysis before it approves major rules. Congress added that mandate in 1996. Weimer, the Pipeline Safety Trust's executive director, said that requirement allows industry to reject fixes it deems too expensive.

A briefing paper issued recently by the Trust points out that a cost-benefit analysis can be effective when costs and benefits can be easily identified and monetized, "but it is not so tidy or easy when trying to value environmental, health and safety factors, as in the pipeline safety field."

"Economic valuation of a healthy child, a clean river, or a safe neighborhood is difficult to undertake," the paper said.

According to research compiled by the Center for Progressive Reform, the Consumer Product Safety Act is the only other federal statute in the health, safety or environment fields that requires a cost-benefit analysis.

Federal rulemaking usually doesn't follow a fixed timeline because it involves so many variables. It tends to be a cumbersome, multi-step process that includes taking the concerns of industry, watchdogs and other interested parties into consideration.

"In general, each stage takes a year," said PHMSA spokesman Damon Hill. The entire process can take "up to five or six years. It depends on all the factors involved."

Access to Spill Response Plans Limited

When the Enbridge accident occurred, local officials in Marshall, Mich., said they knew almost nothing about the pipelines that snaked through their community and weren't prepared for an oil spill of such magnitude.

The NTSB report found flaws with Enbridge's spill response plan and criticized PHMSA for not reviewing the plans more carefully. It pointed out that the agency had just 1.5 full-time positions to manage 450 response plans when the spill occurred in July 2010. The report also noted that both the U.S. Coast Guard and the Environmental Protection Agency have more rigorous review procedures and more staffers to handle the task.

"It is doubtful that the Enbridge plan could have received more than a cursory review," NTSB investigators wrote in their July 10 report. "If PHMSA had dedicated the resources necessary to conduct thorough reviews, it likely would have identified deficiencies and disapproved the Enbridge plan because it lacked sufficient resources for response to a worst-case discharge."

In its Pipeline Safety Act, Congress directed PHMSA to provide copies of all spill response plans to the public upon request, minus any proprietary or security-sensitive information.

But they are still not easy to access. When InsideClimate News recently asked for copies for the existing Keystone pipeline, the proposed Keystone XL and the new Pipeline 6B in Michigan, PHMSA said they would only be available if a request was filed via the Freedom of Information Act.

Spill response plans require pipeline operators to identify personnel and equipment capable of resolving a worst-case oil discharge; pre-position those resources so they can respond efficiently to an emergency; detail a chain of authority for incident response; and describe training, testing and drilling procedures.

The Pipeline Safety Trust is urging PHMSA to make spill response plans broadly available, without the public having to make a special request. The Trust also wants the public to be allowed to review the plans and any revisions and also provide suggestions.

Spill Reporting Still Lax

Pipeline operators are required to report spills to the National Response Center, which is responsible for quickly alerting state and federal agencies about unfolding disasters. Enbridge didn't notify the National Response Center until almost two hours after the company had confirmed the Michigan spill. But that apparently complied with PHMSA's requirement that operators notify the center at "the earliest practicable moment" after spills that caused death, a fire or explosion, significant property damage or water pollution.

The 2012 legislation directs PHMSA to revise that regulation and require pipeline operators to report large spills "at the earliest practicable moment following confirmed discovery" but "not later than 1 hour following the time of such confirmed discovery."

PHMSA has not yet taken steps to revise the rule to meet that directive, said Hill, the agency spokesman.

Slow Progress on Detecting Leaks

The NTSB investigation into the Kalamazoo spill found that Enbridge's leak detection system was partly to blame for the 17-hour gap between when the pipeline ruptured in Michigan and when Enbridge became aware of the spill.

The Pipeline Safety Act calls for the Department of Transportation to spend two years studying the technical limitations of leak detection systems, and to determine if it would be practical to create performance-based standards for those systems.

Ironically, the legislation will delay any new rules in this area, said the NRDC's Swift.

PHMSA had started a rulemaking process to tighten leak detection regulations before the pipeline safety bill was passed. But because the legislation requires PHMSA to spend two years researching the subject, the agency won't be able to issue a final rule until 2014.

"It basically stopped new [leak detection] regulations from coming from PHMSA for at least two years," Swift said. "Given the context of 2011, that's very unfortunate because national attention was on pipeline safety and the need for stronger regulations."

EXTRA: A DILBIT PRIMER

When emergency responders rushed to Marshall, Mich. on July 26, 2010, they found that the Kalamazoo River had been blackened by more than 1 million gallons of oil. They didn't discover until more than a week later that the ruptured pipeline had been carrying diluted bitumen, also known as dilbit, from Canada's tar sands region. Cleaning it up would challenge them in ways they had never imagined. Instead of taking a couple of months, as they originally expected, nearly two years later the job still isn't complete.

Dilbit is harder to remove from waterways than the typical light crude oil—often called conventional crude—that has historically been used as an energy source.

While most conventional oils float on water, much of the dilbit sank beneath the surface. Submerged oil is significantly harder to clean up than floating oil: A large amount of oil remains in the riverbed near Marshall, and the cleanup is expected to continue through the end of 2012.

InsideClimate News spent seven months investigating what made the Marshall spill different from conventional oil spills. Part of the challenge was that there has been little scientific research on dilbit; most of the studies that have been done were conducted by industry and considered proprietary information.

The information we did find comes from government records and publicly available industry studies, plus dozens of interviews with industry analysts, federal and state officials, and several university researchers who've worked with the oil industry. We also interviewed watchdog groups that have focused on increasing dilbit regulations, including the Pipeline Safety Trust, the Natural Resources Defense Council and the Pembina Institute, a well-respected Canadian think tank that supports sustainable energy.

Experts at the University of Alberta and the University of Calgary, where tar sands research has been done, did not return requests for comment. InsideClimate asked the American Petroleum Institute and the Canadian Association of Petroleum Producers to put us in contact with their experts, but neither organization provided scientists or engineers for interviews.

What is dilbit?

Dilbit stands for diluted bitumen.

Bitumen is a kind of crude oil found in natural oil sands deposits—it's the heaviest crude oil used today. The oil sands, also known as tar sands, contain a mixture of sand, water and oily bitumen. The tar sands region of Alberta, Canada is the third largest petroleum reserve in the world.

What makes bitumen different from regular or conventional oil?

Conventional crude oil is a liquid that can be pumped from underground deposits. It is then shipped by pipeline to refineries where it's processed into gasoline, diesel and other fuels.

Bitumen is too thick to be pumped from the ground or through pipelines. Instead, the heavy tar-like substance must be mined or extracted by injecting steam into the ground. The extracted bitumen has the consistency of peanut butter and requires extra processing before it can be delivered to a refinery.

There are two ways to process the bitumen.

Some tar sands producers use on-site upgrading facilities to turn the bitumen into synthetic crude, which is similar to conventional crude oil. Other producers dilute the bitumen using either conventional light crude or a cocktail of natural gas liquids.

The resulting diluted bitumen, or dilbit, has the consistency of conventional crude and can be pumped through pipelines.

What chemicals are added to dilute the bitumen?

The exact composition of these chemicals, collectively called diluents, is considered a trade secret. The diluents vary depending on the particular type of dilbit being produced. The mixture often includes benzene, a known human carcinogen.

If dilbit has the consistency of regular crude, why did it sink during the Marshall spill?

The dilbit that spilled in Marshall was composed of 70% bitumen and

30% diluents. Although the dilbit initially floated on water after pipeline 6B split open, it soon began separating into its different components.

Most of the diluents evaporated into the atmosphere, leaving behind the heavy bitumen, which sank under water.

According to documents released by the National Transportation Safety Board—a federal agency that is investigating the spill—it took nine days for most of the diluents to evaporate or dissolve into the water.

Can conventional crude oil also sink in water?

Yes, but to a much smaller extent.

Every type of crude oil is made up of hundreds of different chemicals, ranging from light, volatile compounds that easily evaporate to heavy compounds that will sink.

The vast majority of the chemicals found in conventional oil are in the middle of the pack—light enough to float but too heavy to gas off into the atmosphere.

Dilbit has very few of these mid-range compounds: instead, the chemicals tend to be either very light (the diluents) or very heavy (the bitumen).

Because bitumen makes up 50 to 70 percent of the composition of dilbit, at least 50 percent of the compounds in dilbit are likely to sink in water, compared with less than 10 percent for most conventional crude oils.

How do you know whether a particular type of crude oil will sink or float?

The industry classifies different crude oils as light, medium or heavy, based on their densities. There is debate over the cutoffs for these categories, but bitumen falls into the "extra heavy" category because it is more dense than water. The diluted bitumen that spilled from 6B was lighter than water and considered heavy crude oil.

But density alone doesn't determine whether a particular type of crude oil will sink or float, said Nancy Kinner, a professor of civil and environmental engineering at the University of New Hampshire who studies submerged oil. Weather and other conditions can change the buoyancy of crude oils: for example, crudes that are lighter than water can sink if they mix with sediment.

That's exactly what happened with the bitumen from 6B. In general, the density of bitumen ranges from slightly heavier than water to barely lighter than water. The bitumen that spilled in Marshall was at the lighter end of the scale. Marc Huot, a technical and policy analyst at the Pembina Institute's Oilsands Program, said the bitumen's density was so close to that of water that it was in "a gray area. It may or may not float depending on [conditions]…think of a log—it floats, but not very well."

But as the bitumen mixed with grains of sand and other particles in the river, the weight of the sediment pulled the bitumen underwater.

Why has it been so hard to clean up submerged oil in the Kalamazoo?

Existing cleanup procedures and equipment are designed to capture floating oil. Because the Marshall accident was the first major spill of dilbit in U.S. waters, cleanup experts at the scene were unprepared for the challenge of submerged oil.

The EPA has supervised the cleanup of nearly 8,400 spills since 1970, but in multiple interviews with InsideClimate News, agency officials said the Marshall spill cleanup was unlike anything they'd ever faced.

"[It's] not something a lot of people have dealt with," said Kinner. "When you can't see [the oil], you don't know where it is, so it's very hard to clean it up."

Once cleanup crews locate submerged oil, it's hard to remove it without destroying the riverbed. Cleanup workers in Marshall were forced to improvise less invasive procedures that balanced oil cleanup with protecting the ecosystem.

On July 16, 2010, just nine days before the Marshall accident, the EPA warned that the proprietary nature of the diluents found in dilbit could complicate cleanup efforts. The agency was commenting on the State Department's Draft Environmental Impact Statement (EIS) of the Keystone XL, a proposed pipeline that would carry Canadian dilbit across six U.S. states and the critically-important Ogallala aquifer.

"First, we note that in order for the bitumen to be transported by the pipeline, it will be either 'diluted with cutter stock (the specific composition of which is proprietary information to each shipper) or an upgrading technology is applied to convert the bitumen to synthetic crude oil,'" the EPA wrote. "…Without more information on the chemical characteristics of the diluent or the synthetic crude, it is difficult to determine the fate and transport of any spilled oil in the aquatic environment.

"For example, the chemical nature of dilutent may have significant implications for response as it may negatively impact the efficacy of traditional floating oil spill response equipment or response strategies. In addition, the Draft EIS addresses oil in general and as explained earlier, it may not be appropriate to assume this bitumen crude/synthetic crude shares the same characteristics as other oils."

How does dilbit affect pipeline safety?

Some watchdog groups contend that dilbit is more corrosive than conventional oil and causes more pipeline leaks. The industry disputes that theory, and there are no independent studies to support either side. In late 2011, Congress passed a bill that ordered the Pipeline and Hazardous Materials Safety Administration (PHMSA) to study if dilbit increases the risk of spills. Results are expected in 2013.

The industry says that Canadian tar sands oil is very similar to conventional heavy crudes from places such as Venezuela, Mexico and Bakersfield, California. Those crude oils, however, aren't transported through the nation's pipelines. The Bakersfield oil is processed at on-site refineries, while the Venezuelan and Mexican imports are shipped via tankers to refineries on the U.S. Gulf Coast.

The same watchdogs that criticize dilbit say that synthetic crude—which is also made from bitumen—poses no additional threats to pipeline safety. The U.S. currently imports more than 1.2 million barrels of Canadian dilbit and synthetic crude per day, and that figure is expected to grow dramatically in the next decade. Most of the increased production will come from dilbit—because Canada's synthetic crude upgraders have reached capacity, and because it's more financially lucrative for U.S. refineries to process dilbit.

Does the government regulate dilbit differently from conventional crude oil?

For the most part, no.

Dilbit is not subject to any additional safety regulations, and PHMSA doesn't track the specific kind of crude oil that flows through each pipeline. This is one of the reasons why it's hard to compare dilbit's safety record with that of conventional crude.

But oil from the tar sands is regulated differently when it comes to taxes. The oil industry pays an 8-cent-per-barrel tax on crude oil produced and imported to the U.S. The tax goes into the Oil Spill Liability Trust Fund, which provides emergency funds for oil spill cleanup and claims. Both the Marshall and BP Gulf Coast spills have tapped that fund.

In early 2011, five months after the Marshall spill, the IRS ruled to exempt dilbit and synthetic crude from paying this tax. According to the energy and environment news service E&E Publishing, the exemption was made "at the request of a company whose identity was kept secret."

Some say the oil from Canada's tar sands is so different based on its chemistry, behavior and how it's produced, that it should not be considered crude oil.

"One would not consider tar sands typical crude," said Kinner, the University of New Hampshire professor. "It's not considered crude oil by most people who deal with oil and oil spills."

Kinner co-directs the Coastal Response Research Center, a collaboration between the university and the National Oceanic and Atmospheric Administration. The center conducts research on innovations in spill response and recently launched a Submerged Oil Working Group.

The tar sands boom is part of a larger industry trend of producing heavier crude oils, whether that's bitumen or conventional heavy crudes, Kinner said. "All the lighter stuff has been used up…we wouldn't be taking tar sands if it wasn't economically viable…and with time, there will be more [spills with] submerged oil."

Anthony Swift, an attorney at the Natural Resources Defense Council who has spent years studying the tar sands industry, said the Marshall spill points to the need for more stringent dilbit regulations.

The Marshall spill is not the largest oil spill in U.S. history, but it is by far the most costly. Using figures from PHMSA's pipeline incident database, Swift calculated that the average cleanup cost of every crude oil spill from the past 10 years was $2,000 per barrel. The Marshall spill has cost upwards of $29,000 per barrel.

"When you have something that isn't the biggest spill we've had, but turns out to be far more damaging and difficult to deal with, that raises the question, what about this spill was different?" Swift said. "And what was different is what spilled."

This primer was written by Lisa Song.
Researcher Lisa Schwartz contributed to this report.

ABOUT THE AUTHORS

ELIZABETH MCGOWAN: Elizabeth H. McGowan was a co-author of *The Dilbit Disaster: Inside the Biggest Oil Spill You've Never Heard Of,* which won the 2013 Pulitzer Prize for National Reporting for InsideClimate News.

McGowan currently covers energy policy for Energy Intelligence in Washington, D.C. She has won multiple awards for her energy and environment reporting at daily newspapers in Wisconsin and Crain Communications. Her freelance work has appeared in a wide array of magazines and other publications.

Her interest in the outdoors and environmental issues was spurred by childhood family camping trips across the United States and Canada as well as ensuing adventures such as a Georgia-to-Maine thru-hike of the 2,167-mile Appalachian Trail and a solo 4,250-mile bicycle ride across the United States. McGowan earned a journalism degree from the University of Missouri-Columbia.

DAVID HASEMYER: InsideClimate News reporter David Hasemyer is co-author of the Dilbit Disaster: Inside the Biggest Oil Spill You've Never Heard Of, which won the 2013 Pulitzer Prize for National Reporting, was a finalist in the 2012 Scripps Howard Awards for Environmental Reporting and won an honorable mention in the 2012 John B. Oakes Award for Distinguished Environmental Journalism. Prior to joining InsideClimate News, he was a freelance journalist whose career included an award-winning tenure at the *San Diego Union-Tribune* as an investigative reporter. Hasemyer's work has been recognized by the Associated Press, the Society for Professional Journalists, the Society of American Business Editors and Writers and the California Newspaper Publishers Association. He has also been a finalist for the Gerald Loeb Award.

Among the articles Hasemyer researched and wrote for the *Union-Tribune* was a series about a 10-million ton pile of nuclear waste, a remnant of the uranium-mining boom in the 1950s and '60s that threatened the Colorado River. Those stories have been widely credited as critical to the U.S. Department of Energy's decision in 2000 to move the pile away from the river. Hasemyer graduated from San Diego State University with a Bachelor's degree in Journalism.

LISA SONG: Lisa Song joined InsideClimate News in January 2011, where she reports on oil sands, pipeline safety and natural gas drilling. She helped write "The Dilbit Disaster" series, which won the 2013 Pulitzer Prize for National Reporting, was a finalist in the 2012 Scripps Howard Awards for Environmental Reporting and won an honorable mention in the 2012 John B. Oakes Award for Distinguished Environmental Journalism. She previously worked as a freelancer, contributing to High Country News, Scientific American and New Scientist. Song has degrees in environmental science and science writing from the Massachusetts Institute of Technology.

INSIDECLIMATE NEWS

InsideClimate News is a Pulitzer prize-winning, non-profit, non-partisan news organization that covers clean energy, carbon energy, nuclear energy and environmental science—plus the territory in between where law, policy and public opinion are shaped. Our mission is to produce clear, objective stories that give the public and decision-makers the information they need to navigate the heat and emotion of climate and energy debates.

We have grown from a founding staff of two to a mature virtual newsroom of ten full time professional journalists and a growing network of contributors. We're aiming to double in size and come to full scale in the next two years.

Climate and energy are defining issues of our time, yet most media outlets are now hard-pressed to devote sufficient resources to environmental and investigative reporting. Our goal is to fill this growing national deficiency and contribute to the accurate public understanding so crucial to the proper functioning of democracy.

INSIDECLIMATE NEWS

HELP KEEP ENVIRONMENTAL JOURNALISM ALIVE

We depend on support from readers like you to produce award-winning journalism to support the public interest.

Please make a fully tax deductible donation to InsideClimate News by visiting our home page and clicking "donate" at the top.

https://insideclimatenews.org

You can subscribe to various newsletters and alerts at no cost from our home page, too.

www.ingramcontent.com/pod-product-compliance
Lightning Source LLC
Chambersburg PA
CBHW070327190526
45169CB00005B/1777